NIMBUS
Technical Development 1934 -59
English edition - 2016

Knud Jørgensen
Ben Geutskens and Richard Reich

NIMBUS
Technical Development 1934 - 59

English edition - 2016

First published in 1988 by Notabene, Copenhagen DK
Danish edition 2005: *Nimbus – teknisk udvikling, 2. udgave* by Motorploven, Hadsten DK

This English edition was published in 2016 by
BoD - Books on Demand GmbH, Copenhagen, Denmark
Cover: Knud Jørgensen

Print: BoD - Books on Demand GmbH, Norderstedt, Germany

ISBN: 978-87-7188-789-1

© 2016 Knud Jørgensen

All rights reserved. With exception of quoting brief passages for the purpose of review, no part of this publication may be recorded, reproduced or transmitted by any means, including photocopying without permission of the author

Contents

Foreword 7
    How to use this book 8

Historical development 9
    Basic construction. 10
    Context. 12
    Patents 13
    Requirement for approval 19

Identification 21
    Year of production 23
    Stock books 24
    Type or model? 26
    Versions 26

Overview of versions 27
    Standard 28
    Standard Extra 30
    Luxus 30
    Sport 33
    Special 33

Colours 36
    Pin striping 39

Numbers etc. 41
    Number plates 47

Overview 52
    1934 52
    1935 54
    1936 55
    1937 57
    1938 60
    1939 62
    1940 – 46 64
    1947 67
    1948 69
    1949 70
    1950 – 51 73
    1952 – 53 75
    1954 – 55 77
    1956 79
    1957 – 59 81

Frame 84
    Centre stand 88
    Foot rests 89
    Knee rubbers 89
    Tool box 89
    Driver – and pillion seat 90

Fuel tank 93
    Fuel valves and –tubes 94

Transfers (Decals) 96

Engine 97
    Cylinder block 98
    Lube oil system 100
    Cylinder head 104
    Valves etc. 106
    Crankshaft 108
    Connecting rods 109
    Pistons etc. 110
    Flywheel 113
    Kick starter 113
    Clutch 114
    Camshaft 116
    Camshaft housing etc. 118
    Oil pan (sump pan) 122
    Exhaust etc. 124

Gearbox	127
Gearshift	132
Clutch release	133
Drive shaft	134
Carburettor	135
Crankcase ventilation	139
Front fork	141
Handlebars	145
Wheels	148
Front wheel	149
Rear wheel	153
Transmission	156
Brakes	157
Front brake	157
Rear brake	159
Mudguards	161
Front mudguard	161
Rear mudguard	163
Speedometer	164
Electrical system	167
Battery	168
Dynamo	169
Regulator	169
Coil	170
Distributor etc.	171
Spark plugs	175
Combination switch	175
Ammeter	176
Charge warning light	177
Horn	177
Brake light switch	178
Head light	179
Tail light	180
Wiring	183
Fuse	185
Instrument illumination	186
Cable clamps etc.	186
Accessories	188
Tool set	189
Tyre pump	189
Military ass.	191
Side cars	192
-Model and version	192
-Chassis	194
-Brakes	198
-Wheels	198
-Mudguards	200
-Lights	201
Overview side cars	201
1934	202
1935	203
1936	204
1937	205
1938	206
1939	207
1940 – 46	209
1947	210
1948 – 51	211
1952 – 53	211
1954 – 56	212
1957 – 59	214
Materials and techniques	215
Surface treatment	218
References	221
Index	222

FOREWORD
English edition 2016

The first edition of 'NIMBUS – teknisk udvikling' was published in 1988 and was reprinted several times. The substantially improved second edition was published in 2005.

This English translated edition with more illustrations is intended for non-Danish speaking readers.

Previously 'NIMBUS – vedligeholdelse', published in 2008, was translated into English and issued in 2012 as 'NIMBUS – Maintenance'.

The purpose of this book is clarified in the chapter below, entitled 'How to use this book'. The content of the book is based upon A/S Fisker & Nielsen's archives, which comprises construction drawings, technical bulletins, sales bulletins, photographs, various brochures, price lists and spare parts lists as well as the manufacturers customer magazine 'Nimbus – Nyt' ('Nimbus –News').

A great number of details of the 1934 model is collected in Jens Bisbjerg Andersen's book 'NIMBUS model C 1934'. The English translation is available on www.geutskens.eu.

The illustrations are in the form of photos, sketches or line drawings. Drawings originated from the factory are marked 'F&N'; all other drawings have been made or modified by the author.

Many thanks to Ben Geutskens, NL, and Richard Reich, GB, for their work on this edition.

Errors or shortcomings remain the author's responsibility.

Højbjerg, Denmark, 31 October 2016

Knud Jørgensen

HOW TO USE THIS BOOK

This book gives an overview of the technical development of the Nimbus-C motorcycle during the production period 1934-1959.

It is a technical handbook as well as a reference book that can be used to study the Nimbus-C motorcycle in detail, to assess condition of the Nimbus-C, and to refurbish or restore the Nimbus-C.
It therefore has an extensive index and in the back a list of key words has been printed. Any part can be found easily by looking it up in the list of key words. In addition, references to other chapters can be found in the text (shown in parentheses).

It is advisable to have the spare parts list readily available whenever the book is being used. Reprints of the spare parts list from 1935, 1938 and 1958 are available. Be aware that the 1958 parts list includes all previous parts lists, but if we deal with a pre 1940 motorcycle, it is more appropriate to use one of the re-printed parts lists from the thirties, because they contain photos that clearly show what a particular part from a certain period should look like.

The text fields of the drawings refer to the range of production numbers during which a particular part was used.

HISTORIC DEVELOPMENT

First of all, a short description of the first Nimbus motorcycles, the models A and B, nicknamed the "Stovepipe".

These first Nimbus models were developed and produced between 1919 and 1927 at Fisker & Nielsen Ltd. Copenhagen, Denmark, by electrical engineer Peder Andersen Fisker.

Motorcycles produced from 1919 – 1923 are called 'Model A', and those from 1924 – 1927 'Model B'.

These models are briefly described here, because some elements from the basic design were also applied to the Nimbus-C. The frame, which gave the motorcycle the nickname »Stovepipe«, was constructed around a sloped large-diameter tube, which had the secondary function of a petrol tank.

Models A and B are fitted with a 746 cc four cylinder in-line engine, with individually casted cylinders. And like the Nimbus-C, the engines are placed in the frame between two flat iron frame rails.

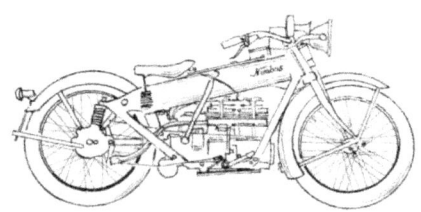

All Nimbuses are shaft driven, but because Models A and B have a sprung rear suspension, this drive shaft is more complicated and is in fact a 'live' drive shaft, whereas with the Nimbus-C, the drive shaft was rigid at first but later provided with a flexible/friction intermediate shaft.

While the front fork constructions for Nimbus-A and -B embodied various types of swing (girder fork) suspension, the Nimbus-C is fitted with a telescopic front fork. The method used (bearings, cones, etc) for supporting the front fork at the headstock is identical for all models. In addition to the mentioned head set parts, the ball on the hand-operated gear lever, and the kick starter pedal are identical parts for Nimbus-A, -B and -C. The tool box is virtually unchanged as are the tool bag, and some tools.

It is worth mentioning that Nimbus motorcycles were fitted with Nilfisk vacuum cleaner parts over the years!

BASIC CONSTRUCTION

Nimbus motorcycles from 1934 until 1959 are called 'Model C or 'Type C'. Throughout this book, we have chosen to use the terms 'Make' for the Nimbus brand, 'Model' for Nimbus-A, -B and -C and 'Version' for 'Standard', 'Standard Extra', 'Luksus/Luxus' and 'Sport'. The Nimbus-C is sometimes referred to as »Nimbus II«. The suffix 'C' is a logical consequence of the fact that Nimbus produced from 1919 – 1923 were called 'Model A' and those produced in the following years, from 1924 – 1928, 'Model B'. Nimbus-A and –B were called 'Stovepipe' by the public.

Immediately after launch, because of its distinctive exhaust sound and the shape of its petrol tank, the Nimbus-C acquired the popular nickname 'Bumblebee'.

Based upon information from the archives, 12,715 Nimbus-C motorcycles were produced. In addition, an unknown smaller number were assembled from spare parts after 1959.

The Nimbus-C was never assembled on an assembly line, but rather in small batches of commonly 25 units. This provided the opportunity to implement many changes along the road. Most of these changes are described in this book, but first of all we will deal with the basic construction, viz. those parts that remained basically unchanged during the years of production:

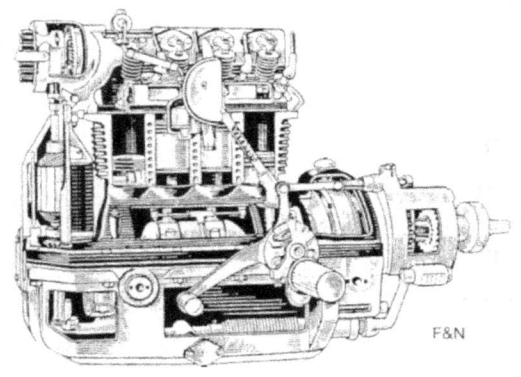

The Nimbus-C is a middle weight motorcycle with a flat-steel frame which is formed around the petrol tank and the engine. The front has been fitted with telescopic forks with front wheel, mudguard and flat steel handlebars, and at the rear, with an unsprung rear wheel with mud- guard. Both wheels are fitted with drum brakes; the front brake is operated by a hand lever, the rear brake by a pedal and a pull rod. The fuel tank has a capacity of 12.1 litre, including 1.5 litre reserve.

The four-stroke engine is a four cylinder in-line engine with overhead valves and cam shaft. The cylinder block is cast in one piece. It has a standard bore of 60 mm, a stroke of 66 mm and a cubic capacity of 746 cc. The pistons are made of aluminum and come in several versions with different dimensions for different compression ratios and power output (this will be discussed later). The engine is air cooled, partly by means of the cooling fins of the cylinder block and head, and partly by the cooling fins of the aluminum crank case, which functions as an oil sump. The aluminum gearbox- and camshaft housing also play a role in the cooling. The overhead camshaft is fitted in a aluminum housing and is driven through the dynamo by means of gear wheels on the Crankshaft. The engine's power is transmitted by a single dry plate clutch, a gear box with three gears and a drive shaft to the pinion and crown wheel of the rear wheel. The lubrication of the Crankshaft, pistons, cam shaft and gear box is done by means of a mechanical oil pump which is driven by the dynamo shaft.

The engine is started by means of a kick starter, which is fitted in the crank case and operated by a pedal arm at the left side of the engine. The carburettor of F&N's own make is of a horizontal flow type, fitted on the intake manifold; it has a vacuum pipe to the cylinder block.

The flat steel handlebars have one twist grip to operate the carburettor (the throttle twist grip) and one to operate the lights (ignition on/off, main beam / low beam). In addition, there are two hand levers, one for the front brake and one for the clutch. The handlebars have a built-in ignition switch and a push button for the horn. The centre stand and the tool box are placed under the engine. On military machines, the toolbox may be placed behind the pillion seat. Furthermore the frame is fitted with rubber knee supports and foot rests.

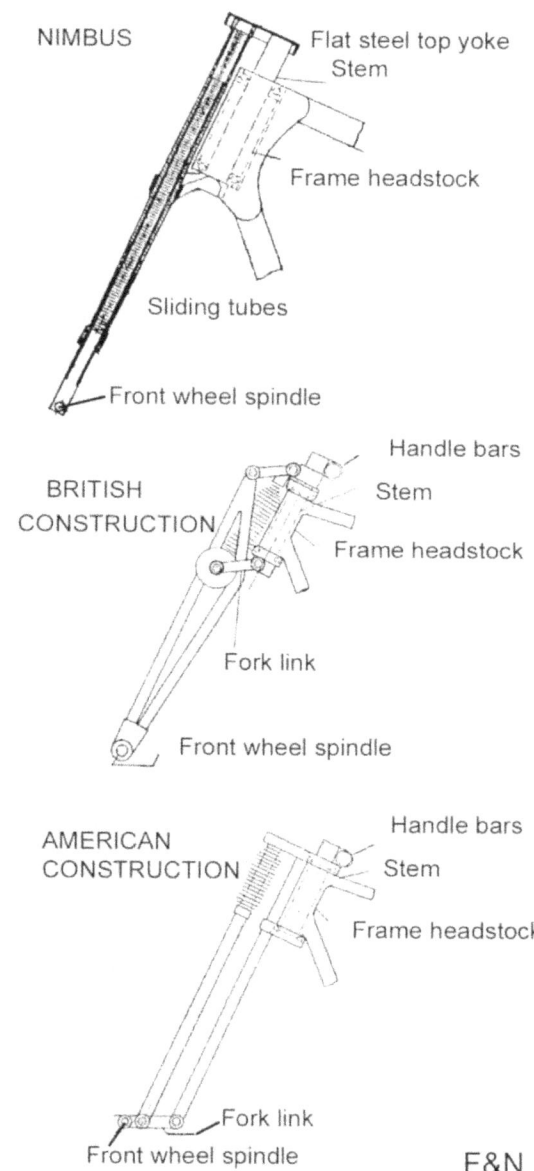

11

The electrical installation is 6 volts and includes a dynamo, a cutter and voltage regulator, and a battery.

It also features a combined ignition and distributor with H.T. leads and spark plugs. Horn, head and taillights are operated from the handlebars, while the brake light, which is fitted in the tail light housing, is operated through a switch activated by the rear brake lever.

CONTEXT

When the Nimbus-C motorcycle was first launched, it was quite naturally compared with other brands. The front forks especially drew a lot of attention.

A comparison between the front fork design principle of the new Nimbus, the British, and American versions respectively is outlined on F&Ns drawings (shown here on page 11).

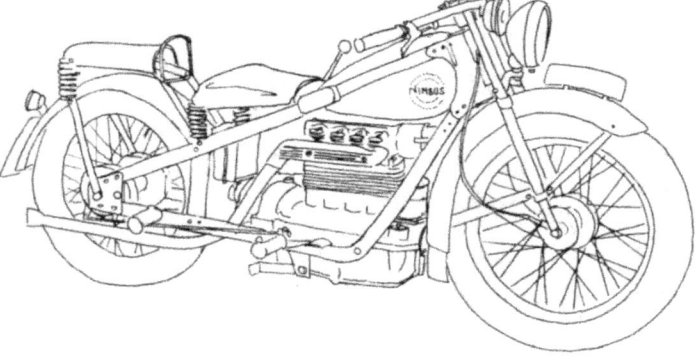

Construction of the front forks for the Nimbus-C can be seen on a 1933 prototype, built by father (F&Ndirector) Peder.Andersen Fisker and his son civil engineer Anders Fisker.

In a number of details, the prototype Nimbus-C differs from final production: the most obvious visual differences are the front mudguard and the tail light, as well as the engine oil filling orifice being on the right hand side.

Furthermore, the prototype shows the exhaust pipe apparently fitting *over* the manifold as opposed to the motorcycles produced later in the series, where the exhaust pipe fits *into* the manifold. In addition, the exhaust manifold of the prototype has no heat shield.

This early machine may, however, be characterized as a 'final version', which was, at the time, far advanced in a considerable number of engineering aspects. There were problems and shortcomings, but these were corrected during the first year, without any cost to the customer.

PATENTS

When the Nimbus-C, later nicknamed »Bumblebee«, was presented to the Danish press on April 20, 1934, A/S Fisker & Nielsen underlined a number of Danish patents related to the construction of this model.

The advantage of having patents is of course protection against imitation and consequently competition in the market, but also to allow third parties to license the patents and pay for them.

Nowadays there are not many Nimbus-C related patents that are applied in other motorcycles, if any.

The patents issued from March – April 1933 are:

Danish patent No. 49451
Motor cycle frame

The flat steel frame is one of the most characteristic features of the Nimbus. It is however not the use of flat steel that is chosen to be patented, but the construction of the frame. The frame embodies the fuel tank and the engine/gearbox and does not just support these. The construction of the frame consequently led to several options that were patented, e.g. the rear mudguard to be hinged and folded back, the position of the centre stand, and the fuel tank fastening. The basic construction of the frame was maintained throughout the years, but the fastening of the headstock varied.

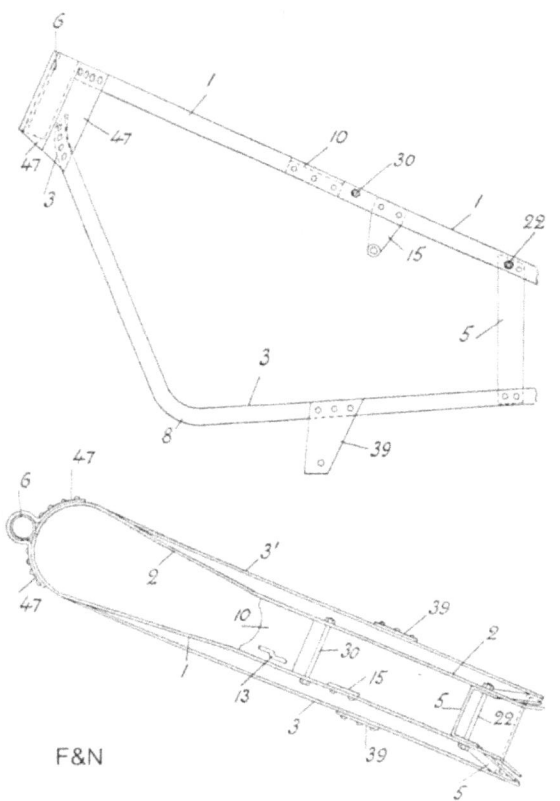

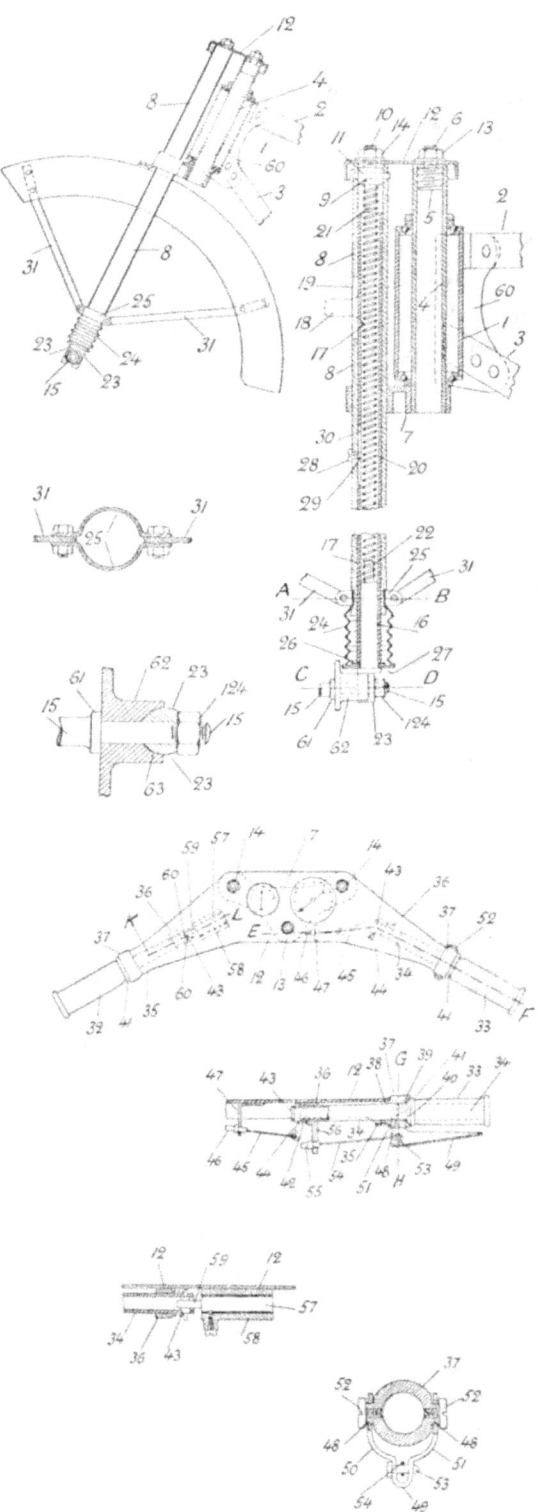

Danish patent No. 49600
Front wheel fork construction for bicycles, especially motorcycles.
The basic principle of a telescopic spring unit is that a tube can move up and down in another tube. The patent includes both the principle of the telescopic spring suspension as well as the attachment of the front wheel to the front forks. The principle of the telescopic suspension was developed at the same time as BMW in Germany, and remained to be used on all later motorcycles. Apart from that, the front fork is one of the Nimbus-C parts that has seen most changes.

Danish patent No. 49189
Handlebars for bicycles, especially motorcycles.
The flat steel handlebars are one of the most characteristic parts of the Nimbus. This patent refers to the construction, and opens a lot of opportunities. Just consider the light switch. A better way of switching the lights on and off cannot be found anywhere. Furthermore, the flat steel handlebars are used as a dashboard for the ammeter and speedometer.
The basic construction of the flat steel handlebars is applied on all later motorcycles, but in time, the outer ends were tilted somewhat and, in combination with the 'high' front forks, the instruments were left out.

Danish patent No. 49734
Construction of the gearbox, especially for motorcycles

Changing gears by means of a hand-change gear lever was most common for motor cyclists in 1934. This patent is only about the simple principle that the gear lever is connected to the gear selector support shaft, by means of an indent in the latter. A foot-change mechanism was introduced as early as 1937, but the hand-change mechanism was not phased out until 1948.

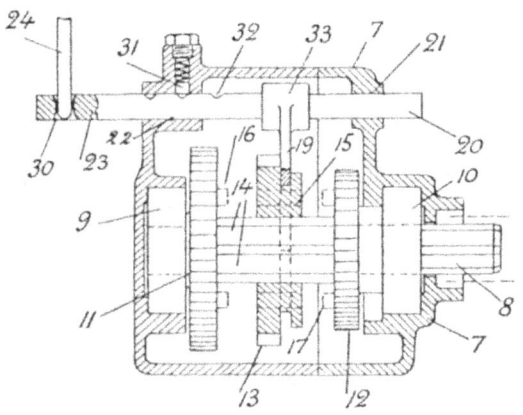

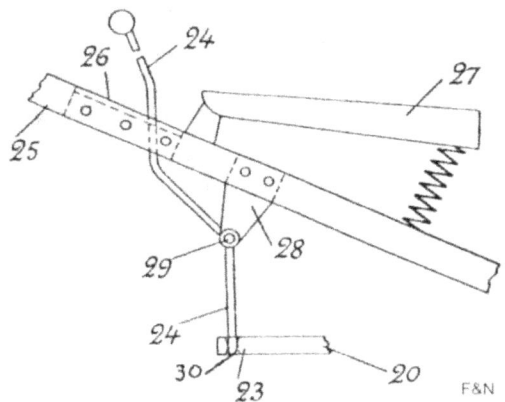

Danish patent No. 48845
Construction of the battery ignition system

This patent is about multiple interrelated aspects, partly about the combined ignition coil and the distributor, partly about the positioning of this combination in the distributor housing by means of a long conductive pin.

This construction has been applied during the whole production period, but the way to limit the rotation of the ignition housing did change.

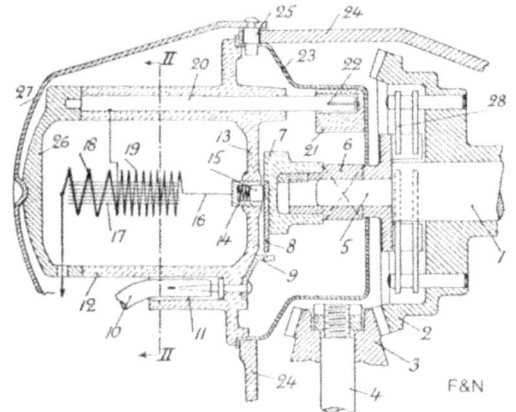

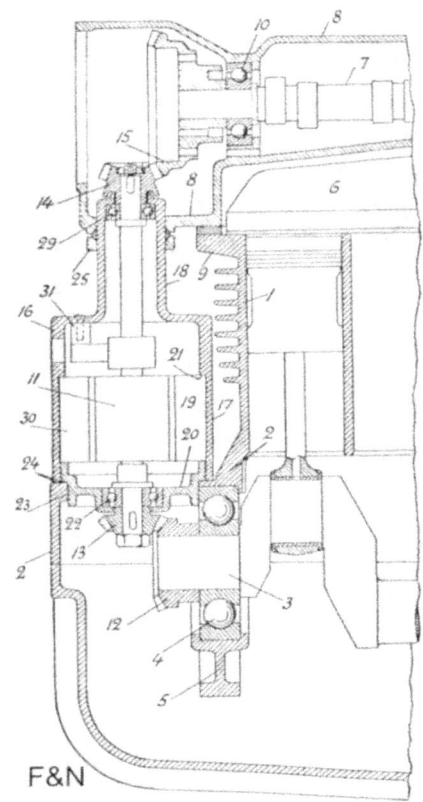

Danish patent No. 49174
Construction of the dynamo-to-engine configuration for motorcycles or similar.

This patent deals with the dynamo being a link in the camshaft propulsion chain, whereby the rotation of the crankshaft is transmitted through the dynamo shaft to the overhead cam shaft. A pinion gear wheel is fitted on either end of the dynamo armature which is engaged with a crown wheel on the crankshaft and the camshaft respectively.

This construction has been used during the entire production period, but the dynamo was almost immediately replaced with different type from the one shown on the drawing. Please note also that the camshaft on the drawing is fitted in ball bearings as opposed to the final construction, where bronze bushes are used as plain bearings. The mechanical oil pump (lube oil pump) is not on the drawing.

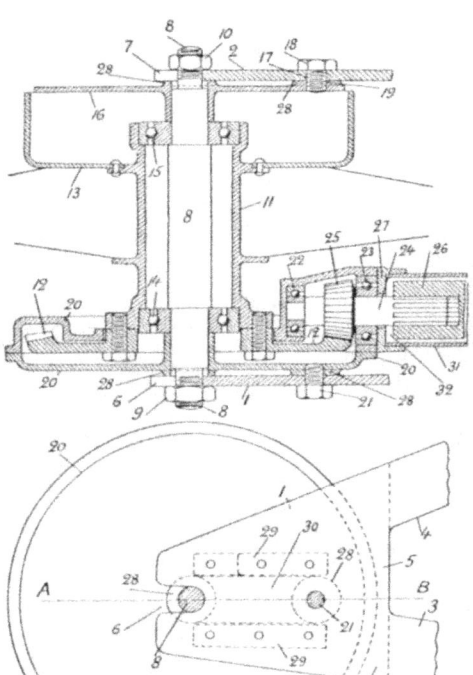

Danish patent No. 48588
Construction of the rear wheel hub and rear frame section for motorcycles.

This patent describes the basic construction that allows the complete rear wheel with transmission and brake to be pulled out of the frame. The two guide rails at either side of the frame, together with the splines of the pinion wheel and drive shaft, require a stiff, unsprung construction. This construction makes changing the rear wheel less difficult.

The basic construction remained throughout the entire production period. Later, the gear wheel housing was provided with two threaded holes instead of one, for improved frame attachment, and in order to obtain better resistance against the driving torque.

Danish patent No. 49358
Construction of engines, especially for motorcycles and similar.
An engine with an overhead camshaft was not very common in 1934, but the principle was well known. The way the rockers were fitted was new however.
This patent describes the way the rockers are supported in the walls of the camshaft housing. The basic construction described was applied to all engines throughout the entire production period, the rocker pivots were however later sealed with covers and gaskets.

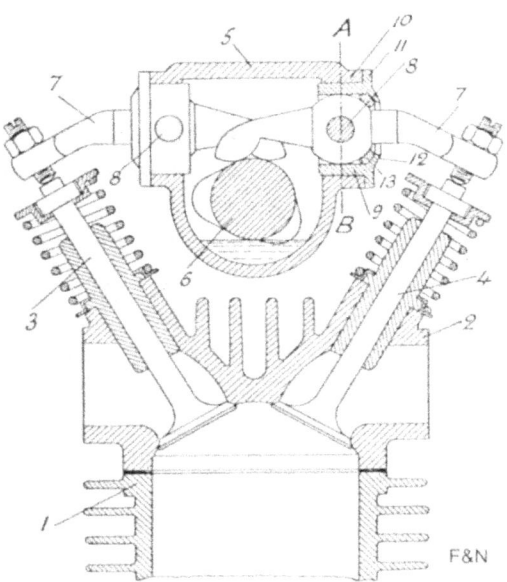

Later additional patents were applied for:
- Suspension of seat and pillion seat
- Rotating valves

Danish patent No. 73411
Seat for bicycles, motorcycles or similar vehicles.
In 1935, the factory used heavy gauge rubber bands for the suspension of sidecar frames; the so-called flat steel frames (see sidecars).

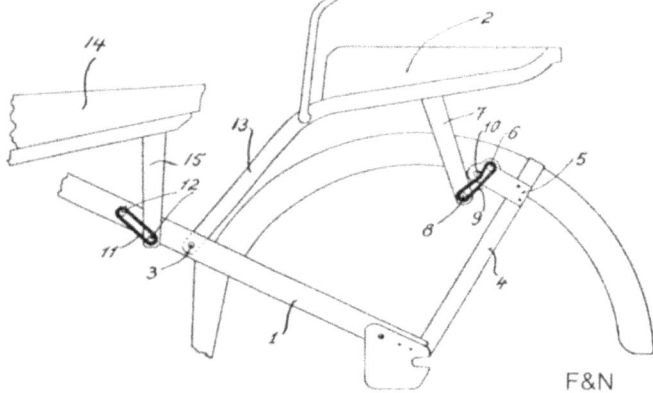

In 1950, from engine no. 9001 onwards, the coil-spring suspension of the seat and the pillion seat were also replaced by rubber bands. The patent was valid as of February 2, 1950 and was applied during the whole production period.
This change is considered to be a major improvement in comfort for both the rider and the passenger for unsprung rear wheels.

Danish patent No. 82279
Valve of the rotating type, especially for combustion engines.
This is perhaps the firm's most technologically ground-breaking patent, which can be directly credited to civil engineer Anders Fisker, and chief tester Anton Marinus Andersen. It is valid as of February 2, 1955.

Even though the patent was developed before the production period started, just a few test engines were built with these valves. A more detailed description of the patent is therefore beyond the scope of this book.

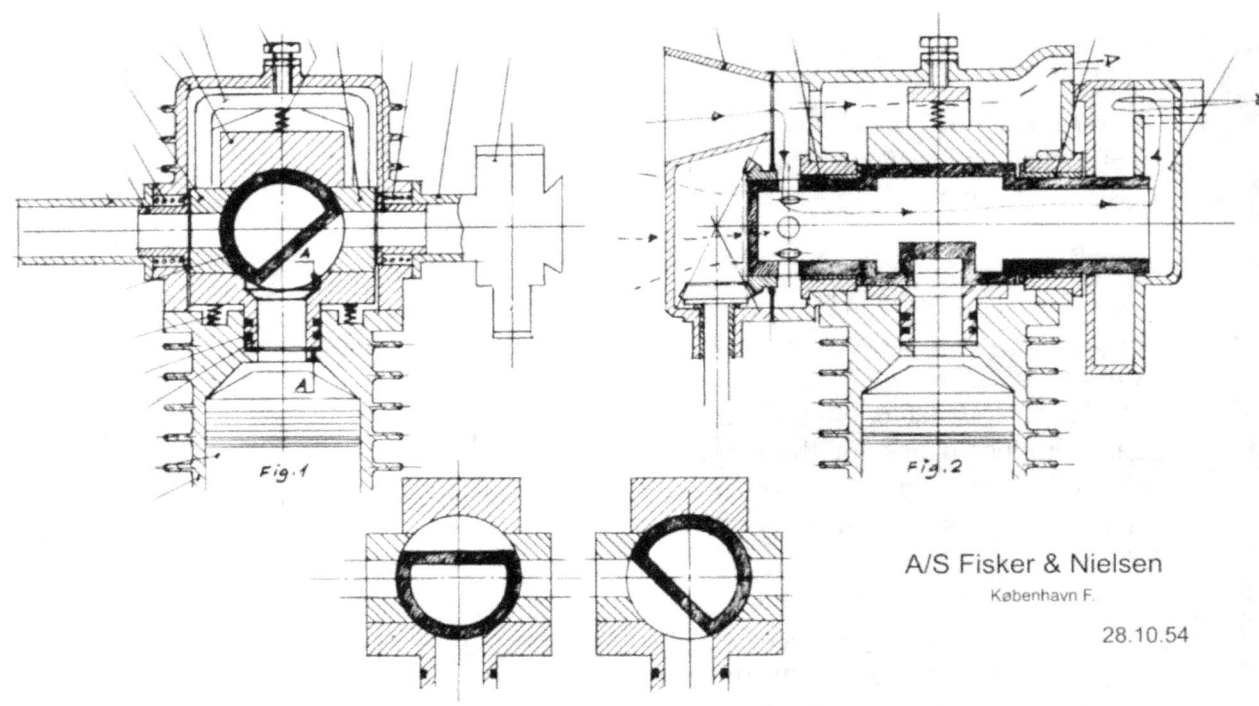

REQUIREMENTS FOR APPROVAL

Denmark introduced a new Road Traffic Act in 1932. At the same time the 1903 Motor Law was changed. The Motor Law describes amongst others the provisions for the approval of vehicles. The Motor Law from 1932 describes the requirements applicable to the Nimbus-C from 1934 – 1955.

Every single new vehicle in Denmark had to be approved for registration and this system remained in place until the new law of 1954 was put in place (applicable as of 1955 – 1956)

The number of new vehicles to be approved, increased dramatically, and it was considered to be unreasonable that every single unit of a batch of identical vehicles had to be tested and approved.

Since then a vehicle is registered based upon the current registration regulations in Denmark.

During the production period of the Nimbus-C, a whole series of changes with respect to test requirements affecting the production took place. The factory was prepared for the first change. A marketing brochure dated November 2, 1934 mentions amongst other things:

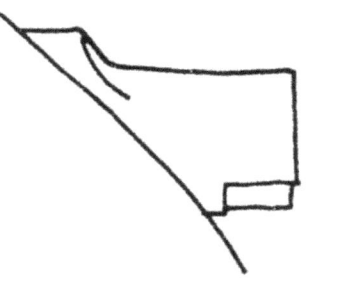 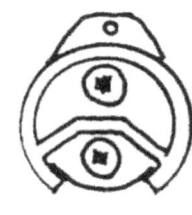 

Brake lights: because of an inquiry related to the brake light regulation, we inform you that this light is already fitted to cope with possible future legislation, so that our motorcycles can quickly be brought in accordance with the law, without any difficulties.

In 1955 the factory again prepared in advance for a new law, which affected the tail light glass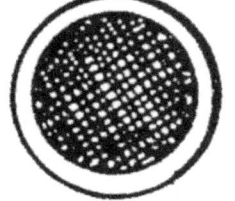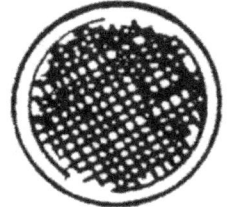
with reflector. But it appeared not to meet the requirements. The new glass had to be replaced after a short time with the 'J.R.U 129' marked tail light glass (see the applicable section).

The current (2016) main rule for approval in Denmark is that a motorcycle has to comply with the law, applicable at the time of first approval. This applies amongst others for lights and brakes. But the consequence can be that the vehicle has a limited approval, e.g. riding only by daylight or with limited speed.

Fortunately, the Nimbus-C is approved for Denmark without restrictions. Any doubt is due to the fact that no reference was made to the original design. We hope that this book may contribute to a flawless approval in other countries.

For the Nimbus-C a standard model approval has been issued, but be aware that it is only valid as of frame number 15001, hence as of April 1, 1956.

For the Nimbus-C sidecar a model approval has been issued for sidecar frame RB from June 15, 1948. The approval for the weight is part of it. The weight of the sidecar alone is 45 kg. The weights of the box or coach and the goods have to be added. The total weight of this load may not be greater than 210 kg. So the maximum allowable weight of vehicle and load must not exceed 255 kg.

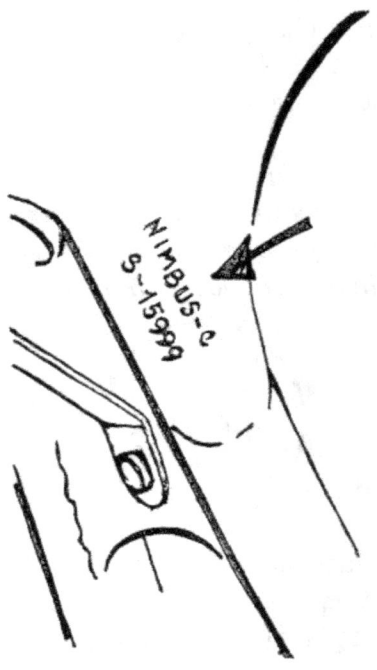

IDENTIFICATION

Determining the year of production, version and colour of the Nimbus-C can be difficult. The starting point is the production number of the motor-cycle.

From 1934 – 1955, this number is the same as the frame- and the engine number. Until March 31, 1956, the frame- and engine number is stamped in various places:
- Between the tank and the seat
- On the handlebars
- At the left hand side headstock support web.
(See the chapter 'Numbers etc.').

The engine number (which is the same as the production number) can be found at the left hand side of the cylinder block, normally straight under the carburettor. During a short period of time around 1940 it was stamped over the foremost engine mounting bolt. Where the engine number differs from the frame number in the sequence 1301 – 13572, at some point in time the engine block has been replaced.

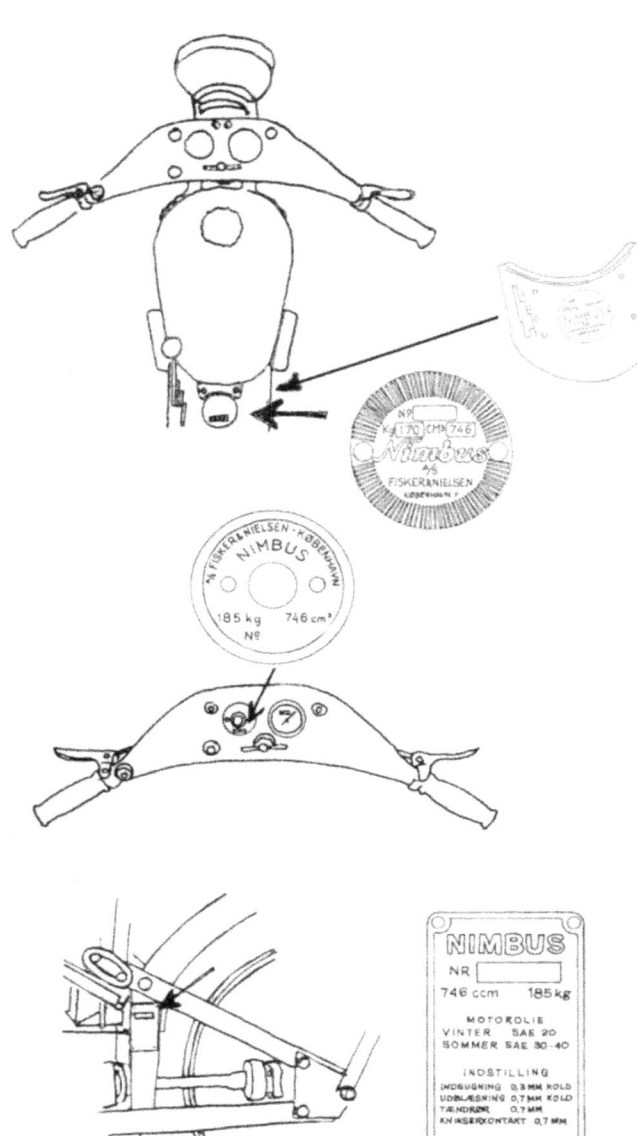

After April 1st 1956, the production number continued to be stamped on the cylinder block as well as on the maker's identification plate under the seat, but the frame's number is now stamped on the frame at the left hand side below the bottom flange of the headstock support web.

NIMBUS-C
S – 15999

Please note that from this point of time the engine- and production numbers are not necessarily identical!

If the characters on the frame reads e.g.

TO 99999

does this mean that the Danish customs authorities, at some point in time, have accepted the replacement of a frame, e.g. because of a severe damage. The TO-numbers cannot be used to determine the year of production.

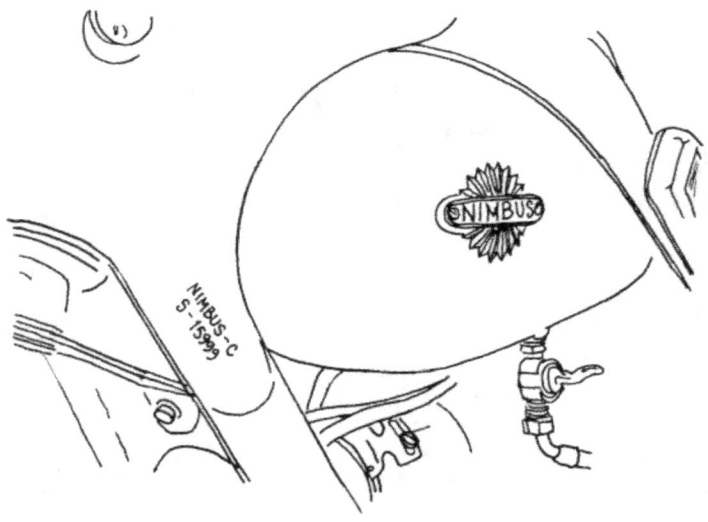

YEAR OF PRODUCTION

There are several ways to determine the year of fabrication of a Nimbus-C
- the year of *manufacture* by the factory
- the year in which the machine was *sold* by the factory
- the year in which the first *registration* took place

All three methods have an uncertainty.

In Circular No. MC 15 from 1935, the company writes to the dealers:

Series as opposed to year of production: Contrary to other motorcycle manufacturers, it has always been our policy to produce in series and not in years, from the point of view that it does not make sense to make customers believe that they get something really new while year-to-year changes are in fact without meaning – or to supply a new model which has not sufficiently been tested under working conditions. Hence, a new Nimbus version will only be launched after substantial changes.

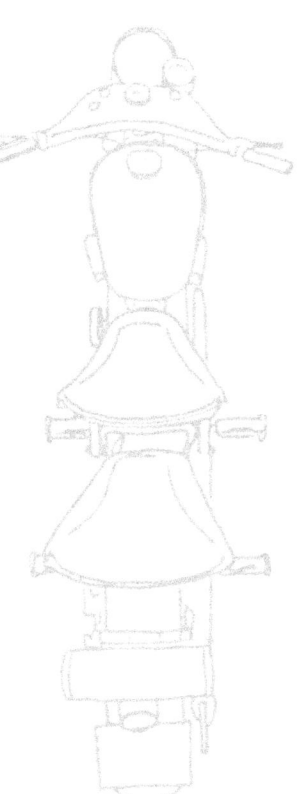

By using the original stock books from A-S Fisker & Nielsen's production and sales of the Nimbus-C a list has been drafted for this book that shows the approximate production numbers per year. Other lists show a mix of 'produced' and 'sold' quantities. From the stock books it cannot be concluded when the machine has been produced, but rather when it was put into stock (therefore it is called a »stock book«). It can be noticed that machines that were in stock in December, were not sold until much later in the following year. This is due to the fact that Nimbus-C was produced in very small series, so machines listed in the stock books with sequential numbers can be of a different version as well as of a different colour. But even the revised list in the stock book is not very precise when it comes to numbers around the end of the year.

This is especially uncertain with numbers issued under the German occupation of Denmark from 1940-45. During this period the systematic in the stock books have completely gone. (This was done on purpose to confuse the occupying force, which showed great interest in the magnitude of the motorcycle production and the spare capacity, if any).

The year in which the machine was sold (called 'ab fabrik') can also be determined by using the stock books, but only with a greater uncertainty than the year of production. As mentioned before, machines in stock per December were sold much later in the following year. Likewise, certain Nimbus-C versions for the following year could already be bought in November or December.

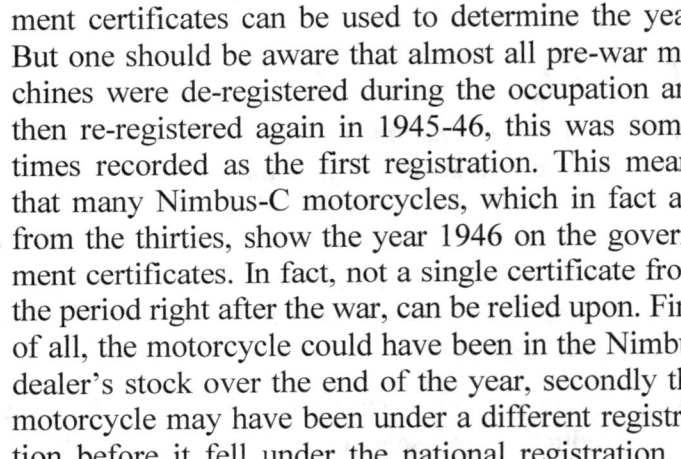

Finally, the section 'First registration' mentioned in the government certificates can be used to determine the year. But one should be aware that almost all pre-war machines were de-registered during the occupation and then re-registered again in 1945-46, this was sometimes recorded as the first registration. This means that many Nimbus-C motorcycles, which in fact are from the thirties, show the year 1946 on the government certificates. In fact, not a single certificate from the period right after the war, can be relied upon. First of all, the motorcycle could have been in the Nimbus dealer's stock over the end of the year, secondly the motorcycle may have been under a different registration before it fell under the national registration. It may e.g. have been a military vehicle, as often was the case, or it may have been registered abroad.

STOCK BOOKS

A/S Fisker & Nielsen's stock books for Nimbus-C motorcycles and sidecar frames can be found in the archives of the marque club *Danmarks Nimbus Touring*.

A digital copy of the stock books for Nimbus-C motorcycles and sidecars, available for members of *Danmarks Nimbus Touring*, can be found on www.nimbus.dk. There are a few differences between the original stock books and the digital copy – misprints which might be corrected once in the future. In case of doubt, the *Danmarks Nimbus Touring* archive can be consulted.

The stock books show the date a machine has been invoiced according to the factory's book-keeping and to which dealer, or otherwise, it was sold. For the numbers 1301 – 11201 the stock books also show the version and colour. The first owner, which was supplied by the dealer, may be stated too. For motorcycles having a number after S-15001, with engine numbers 13573 – 14015 information relating to the number of the key for the steering lock is listed.

YEAR OF PRODUCTION FOR NIMBUS-C

Year		
	1934	1301 - 1500
-	1935	1501 - 2014
-	1936	2015 - 2646
-	1937	2647 - 3489
-	1938	3490 - 4426
-	1939	4427 - 5512
-	1940 - 45	5513 - 6150
-	1945	6151 - 6406
-	1946	6407 - 7064
-	1947	7065 - 7500
-	1948	7501 - 8000
-	1949	8001 - 8825
-	1950	8826 - 9704
-	1951	9705 -10399
-	1952	10400-11420
-	1953	11421-12178
		12180-12190
		12212-12223
-	1954	12179
		12191-12211
		12224-13009
-	1955	13010-13572
-	1956	13573-13769
		13775-13777
		13801-13854
-	1957	13770-13774
		13778-13800
		13855-13900
-	1958	13901-13953
-	1959	13954-14015

TYPE or MODEL?

The use of terminology *type* and *model* has not been used by the factory consistently!
Until 1935, this was not a problem, but even though at this point of time different motorcycle designs hit the market, the term *model* was both used for the motorcycle itself as well as for the different designs.

NIMBUS is the term for the *make* of this motorcycle.
> Nimbus 1934 – 59 is the term for the approved motorcycle *model* Nimbus-C.
> Nimbus-C has been produced in several different *versions*.

Therefore, throughout this book the terminology *model* C is used for the Nimbus-C and the different designs of the Nimbus-C will be called *versions*.

VERSIONS

The factory used the following terminology for the different Nimbus designs:
'Standard', 'Standard-Extra', De Luxe, 'Sport' and 'Special.

Some dealers fitted additional accessories to the motorcycle and (for example) would rename the version from 'De Luxe' to

»Travel« even though the factory frowned on such initiatives. In Circular MC. 29, from October 1936:

" *Finally we request dealers not to advertise Nimbus in different versions other than the original versions, supplied by the factory.*"

In the stock books, the original version of every single machine from 1301 until 11200 can be found. We can assume that all 2815 machines after 11200, supplied to the postal services and the army, were 'Standard' vehicles. The remaining machines were either 'Standard' or Special, but unfortunately, the colour is not listed.

The registration documents may show a version, but it cannot be relied upon that this is correct, especially not, in the case where the registration certificate has been renewed. It may be interesting to investigate this through former owners or their descendants, but this may not be very reliable, unless their memories can be supported by clear pictures from the early years.

OVERVIEW OF VERSIONS

Overview of Nimbus-C versions and their characteristics:

The comparison includes the following:

1. Enameling of the frame, front fork, handlebars, top of fuel tank, lamp brackets where applicable and mudguards.
2. Striping of the fuel tank and the mudguards (see chapter *Striping*).
3. Enameling of the lamp housing, horn (with or without front plate), tool box, centre stand, seat- and pillion seat frame and springs.
4. Surface treatment of the lower part of the fuel tank: Chrome plated or aluminum enameled.
5. Surface treatment of the hubs, brake drums and – anchor plates and final drive housing: Chrome plated or aluminum enameled.
6. Surface treatment of the spokes, nipples and rims.
7. Where surface treatment of the drive shaft, foot-change gear pedal (where applicable), brake- and clutch pedal

(where applicable), their pull rods, kick-start pedal and clutch release lever, are called 'electro-enameled', this appears to be sprayed with aluminum- coloured enamel (see chapter *Colours*). This is true for all versions. All brake levers are equal and 'electro-enameled'.
8. Polishing, if any, of aluminum parts, like oil pan and brake anchor plates.
9. Surface treatment of exhaust pipe.

The comparison does *not* include:
1. The fuel tank lid, fuel line, lube oil suction pipe, choke / air filter, brake- and clutch levers, hand-change gear lever, exhaust heat shield, speedometer ring, bracket for ignition coil, pillion seat handgrip, oil return pipe from camshaft housing, keyhole plate, horn front plate (if any) and several small parts.
2. Valve spring cups, valve springs, rocker supports, distributor housing, oil filter flange, fuel line pillars, bolts, nuts, fittings and mountings. These earlier parts are nickel plated, later chrome- or cadmium plated.
3. Fork tubes for 'high' front forks and lamp housing rings. These are nickel- or chrome plated for civil motorcycles and enameled for military machines.

Finally, the overview lists special information for each version.

'STANDARD'
This name is used for the ordinary machine for everyday use. This is not the case for the years 1946-53, where all machines were called *"Special"*, irrespective their use.

'STANDARD'-1
1934
1. Frame, front fork, handlebars, top of fuel tank and mudguards: Black enamel
2. No striping
3. Lamp housing, horn, toolbox, centre stand, seat- and pillion seat frame and
springs: Black enamel.
4. Lower part of fuel tank: Aluminum enamel.

5. Brake drums and –anchor plates and final drive housing: Black enamel.
6. Hubs, spokes, nipples, rims: Black enamel.
7. Drive shaft, foot rests, brake- and clutch pedal and their pull rods, kick-start lever, clutch release lever: Black enamel.
8. Oil pan: Unpolished.
9. Exhaust pipe: Black enamel.

1935-45

1. Frame, front fork, handlebars, top of fuel tank and mudguards: Black enamel.
2. No striping.
3. Lamp housing, horn, toolbox, centre stand, seat and pillion seat frame and springs: Black enamel.
4. Lower part of fuel tank: Aluminum enamel.
5. Hubs, brake drums and – anchor plates and final drive housing: Black enamel.
6. Spokes, nipples, rims: Black enamel.
7. Drive shaft: Black enamel. Foot rests, foot-change gear pedal (where applicable), brake pedal, clutch pedal (where applicable), their pull rods, kick-start pedal and clutch release lever: Aluminum enamel.
8. Oil pan: Unpolished.
9. Exhaust pipe: Black enamel.

'STANDARD'-2
1954-59

1. Frame, front fork, handlebars, top of fuel tank, lamp brackets and mudguards: Black enamel.
2. Double gold striping
3. Lamp housing, horn, front plate included, toolbox, centre stand, seat and pillion seat frame: Black enamel.
4. Lower part of fuel tank: Aluminum enamel.
5. Hubs, brake drums and -anchor plates and final drive housing: Aluminum enameled.
6. Spokes and nipples: Cadmium plated. Rims: Aluminum

enameled.
7. Drive shaft, foot rests, foot-changed gear pedal, brake pedal, their pull rods, kick-start pedal, clutch release lever: Aluminum enamel.
8. Oil pan: Unpolished.
9. Exhaust pipe with fish tail: Black enamel.

'STANDARD-EXTRA'
1936
1. Frame, front fork, handlebars, top of fuel tank and mudguards: Black, red or green enamel.
2. Single gold striping.
3. Lamp housing, horn, toolbox, centre stand, seat- and pillion seat frame and -springs: Black enamel.
4. Lower part of fuel tank: Aluminum enamel.
5. Hubs, brake drums and –anchor plates, final drive housing: Black, red or green enamel.
6. Spokes, nipples and rims: Aluminum enamel.
7. Drive shaft, foot rests, foot-change gear pedal (where applicable), brake- and clutch pedal, their pull rods, kick-start pedal and clutch release lever: Aluminum enamel.
8. Oil pan: Unpolished.
9. Exhaust pipe: Black enamel.

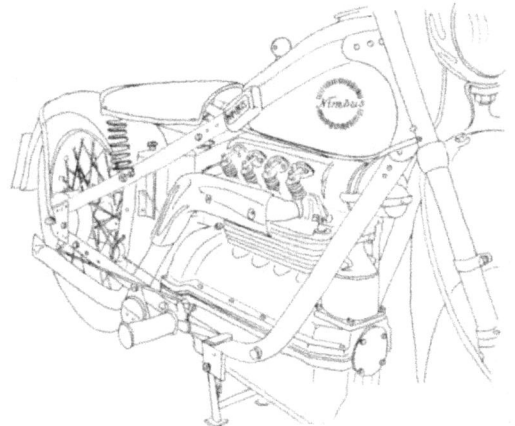

'LUXUS' - 1
1935
1. Frame, front fork, handlebars, top of fuel tank and mudguards: Black, red or green enamel.
2. Single gold striping.

3. Lamp housing, horn, toolbox, centre stand, seat- and pillion seat frame and -springs: Black enamel.
4. Lower part of fuel tank: Chrome plated.
5. Hubs, brake drums and –anchor plates and final drive housing: Black, red or green enamel.
6. Spokes, nipples and rims: Chrome plated.
7. Drive shaft, foot rests, brake- and clutch pedal, their pull rods, kick-start pedal and clutch release lever: Aluminum enamel.
8. Oil pan: Polished.
9. Exhaust pipe: Chrome plated.

1936

1. Frame, front fork, handlebars, top of fuel tank and mudguards: Black, red or green enamel.
2. Single gold striping.
3. Lamp housing, horn, toolbox, centre stand, seat- and pillion seat frame and -springs: Black enamel.
4. Lower part of petrol tank: Chrome plated.
5. Hubs, brake drums and –anchor plates and final drive housing: Aluminum enamel.
6. Spokes: Cadmium plated. Nipples and rims: Chrome plated.

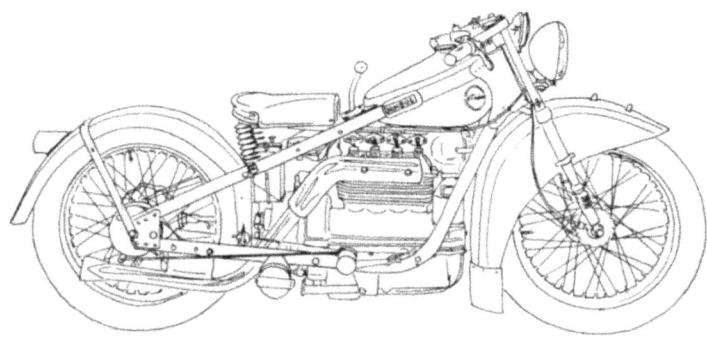

7. Drive shaft, foot rests, brake- and clutch pedal, their pull rods, kick-start pedal and clutch release lever: Aluminum enamel.
8. Oil pan: Polished.
9. Exhaust pipe: Chrome plated.

1937-45

1. Frame, front fork, handlebars, front fork, top of fuel tank and mudguards: Red or green enamel.
2. Single gold striping.
3. Lamp housing, horn, toolbox, centre stand, seat- and pillion seat frame and -springs: Black enamel.

4. Lower part of fuel tank: Chrome plated.
5. Hubs, brake drums and –anchor plates and final drive housing: Aluminum enamel.
6. Spokes, nipples and rims: Aluminum enamel.
7. Drive shaft, foot rests, foot-change gear pedal (where applicable), brake- and clutch pedal, their pull rods, kick-start pedal and clutch release lever: Aluminum enamel.
8. Oil pan: polished.
9. Exhaust pipe: Chrome plated.

'LUXUS' - 2
1954-59
1. Frame, front fork, handlebars, top of fuel tank, lamp brackets and mudguards: Black, red, withered green or deep sea green enamel.
2. Double gold striping.
3. Lamp housing, horn excl. front plate, tool box, centre stand, seat- and pillion seat frame: Same as frame etc.
4. Lower part of fuel tank: Chrome plated with enameled with Nimbus plates.
5. Hubs and final drive housing: Aluminum enamel. Brake drums and –anchor plates: Polished.
6. Spokes: Cadmium plated. Nipples and rims: Chrome plated.
7. Drive shaft, foot rests, foot-change gear pedal, brake pedal, its pull rod, kick-start pedal and clutch release lever: Aluminum enameled.
8. Oil pan: polished.
9. Exhaust pipe with fish tail: Chrome plated.

'SPORT'
1937-47

1. Frame, front fork, handlebars, top of fuel tank and mudguards: Blue enamel.
2. Single silver striping.
3. Lamp housing, horn, toolbox, centre stand, seat- and pillion seat frame and -springs: Black enamel.
4. Lower part of fuel tank: Chrome plated.
5. Hubs, brake drums and –anchor plates and final drive housing: Aluminum enamel.
6. Spokes, nipples and rims: Chrome plated.
7. Drive shaft, foot rests, foot-change gear pedal, brake pedal, its pull rod, kick-start pedal and clutch release lever: Aluminum enamel.
8. Oil pan: Polished.
9. Exhaust pipe with large fish tail, fitted with brackets: Chrome plated. This is typical for the 'Sport' version.

Other characteristic features for the 'Sport' version are: High compression engine with foot-change gear. Off-the-road rear tyre 3.50" x 19" and lengthways corrugated front tyre, 3.25" x 19".

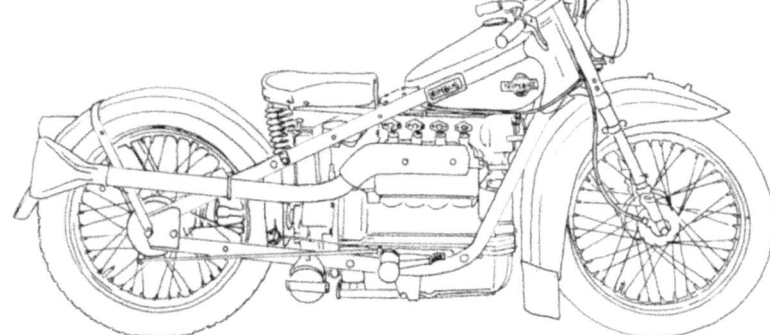

'SPECIAL' -1
1939-45

1. Frame, front fork, handlebars, top of fuel tank and mudguards: Ivory/yellow or lavender/grey (as of 1943: polychromatic grey).
2. Single gold striping for ivory/yellow and lavender/grey.
3. Lamp housing, horn, toolbox, centre stand and seat- and pillion seat frame and springs: Black enamel.
4. Lower part of fuel tank: Chrome plated.
5. Hubs, brake drums and –anchor plates and final drive

housing: Aluminum enamel.
6. Spokes and nipples: Cadmium plated. Rims: Chrome plated.
7. Drive shaft, foot rests, foot-change gear pedal (where applicable), brake pedal and clutch pedal (where applicable), their pull rods, kick-start pedal and clutch release lever: Aluminum enamel.
8. Oil pan: Polished.
9. Exhaust pipe: Chrome plated.

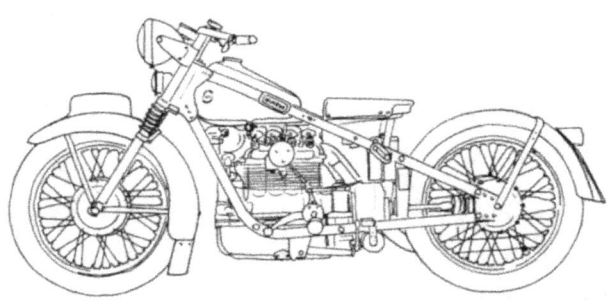

Characteristic features for the 'Special' version are: High compression engine with foot- or hand-change gear at choice. Off-the-road rear tyre 3.50" x 19" and lengthways corrugated front tyre, 3.25" x 19".

'SPECIAL' -2
1946-52

1. Frame, front fork, handlebars, top of fuel tank and mudguards: Blue or black enamel (1950 – 52 black only).
2. Single silver striping for blue and single gold striping for black.
3. Lamp housing, horn, toolbox, centre stand and seat- and pillion, seat frame and -springs: Black.
4. Lower part of fuel tank: Chrome plated or aluminum enamel.
5. Hubs, brake drums and – anchor plates and final drive housing: Aluminum enamel.
6. Spokes and nipples: Cadmium plated. Rims: Chrome plated or aluminum enamel.
7. Drive shaft, foot rests, foot-change gear pedal (where applicable), brake pedal, their pull rods, kick-start pedal, clutch release lever: Aluminum enamel.
8. Oil pan: Polished.
9. Downwards bent exhaust pipe with fish tail, fitted with brackets: Black enamel.

1952-53

The same as 1946-52, but in black or red and with reduced compression (see pistons) and with double gold striping.
This version should consequently have been called 'Standard', compared with 'Standard' 1954-59 above.

Enameled parts produced by Fisker & Nielsen Ltd. are bake enameled. The enamel of parts from third parties, like lamp housing and horn, are unbaked.

NIMBUS-C – VERSIONS

	'Standard'	Standard - Extra	'Luxus'	'Sport'	'Special'
1934	St-1				
1935	St-1		Lu-1		
1936	St-1	SE	Lu-1		
1937	St-1		Lu-1	Spo	
1938	St-1		Lu-1	Spo	
1939	St-1		Lu-1	Spo	Spc-1
1939-44	St-1		Lu-1	Spo	Spc-1
1945	St-1		Lu-1	Spo	Spc-1
1946				Spo	Spc-2
1947				Spo	Spc-2
1948					Spc-2
1949					Spc-2
1950					Spc-2
1951					Spc-2
1952					Spc-2
1953					Spc-2
1954	St-2		Lu-2		
1955	St-2		Lu-2		
1956	St-2		Lu-2		
1957	St-2		Lu-2		
1958	St-2		Lu-2		
1959	St-2		Lu-2		

COLOURS

The determination of the original colour of the Nimbus-C can be done in various ways.

The factory stock books show in which colour Nimbus-C were delivered for production numbers from 1301-11200.
For Nimbus-C motorcycles after 11200, the original colour can only be determined for machines used for military-, police- or mail service.

Individual machines may have been resprayed before sale. In this case the description of the version is available, e.g. from the contract or bill of sale. The colour table will show the correct colour.

Finally, one can look for remainders of the original enamel, still present on the motorcycle. This will of course only make sense if the motorcycle has not been totally sandblasted! Usually, the factory coating consists of a brown primer as a base coat, followed by a red filler primer, on top of that a grey top coat and finally the coloured enamel. Remainders of the original enamel are most likely to be found at the underside of the handlebars and the inside of the headstock.

Originally, Nimbus-C was given a chemical surface treatment, parkerising or bonderising (an anti-corrosive phosphating process), followed by a varnish coating from Danish paint- and varnish factories. During the whole production period, enameling with cellulose varnish was applied, which could easily be applied for spraying or dipping. It had a reputation for brightness, strength and excellent bonding. Later, this type of varnish was replaced by full synthetic enamel. Today, most people will prefer two-component enamel, which is very durable. In some cases, it may be advantageous to use powder-coating for wheels and frame (see the chapter on *surface treatment*)

There are several standard colours for Nimbus-C, as can be seen from the colour scheme below, but it is important to realize that the factory was, upon request, willing to supply the Nimbus-C in any colour. Besides the standard colours as listed in the stock books, machines have been supplied in white, brown, silver-grey, light-blue or cream as well as fully chrome plated! In addition, quite a few machines were either supplied in, or re-enameled to a firm's brand-colour. Some standard colours have changed over the years like 'Danish postal-services yellow' that became darker over the years and not just because the old enamel had faded! The army colour has also changed many times over the years, but if you want to be a perfectionist you need to use sources outside this book.

NIMBUS COLOURS

The author of this book has previously tried to find the individual enamel recipes, but because the colour names may have changed, and the technology of the pigments evolved, it was decided to use RAL-codes in this book instead. There are certainly other colour systems besides the RAL method, but in the field of automobile enamels, the RAL system is still (2016) considered to be unambiguous. (See http://www.ralcolor.com/).

The overview below is highly simplified and all colours (but one) can be defined in the RAL colour system. However, this means that some shades may be slightly different from the original.

Nimbus Paint Colours - and Modern Equivalents

Colour	Colour Name Original name	Colour Name RAL system	RAL-Number	Year Used
Black 1	Black	Jet Black	RAL-9005	1934-60
Red 6	Red	Wine Red	RAL-3005	1935-45
Red 7	Bordeaux	Purple Red	RAL-3004	1952-59
Green 2	Green	Pine Green	RAL-6028	1935-45
Green 3	Withered Green	Reseda Green	RAL-6011	1954-60
Green 4	Deep Sea Green	Blue Green	RAL-6004	1954-60
Blue 13	Tivoli Blue	Steel Blue	RAL-5011	1937-45
Blue 14	Blue	Green Blue	RAL-5001	1946-50
Yellow 8	Ivory/Yellow	Ivory	RAL-1014	1939-45
Grey 10	Lavender/Grey	Grey Beige	RAL-1019	1939-43
Grey 11	Polychromatic Grey	Grey Aluminum	RAL-9007	1943-45
Other	**Defence Colours**			
Green 5a	Olive/Grey	Olive Grey	RAL-7002	1934-45
Green 5b	Grey/Olive	Grey Olive (matt)	RAL-6006	1945-60
Grey 12	Civil defence Grey	Traffic Grey A	RAL-7042	1950-60
Yellow 9	Danish Post Yellow replaced by	Maize Yellow	RAL-1006	1934-60
Electrostatic Enamel Bright Silver (used on the lower part of the fuel tank and other misc parts)		White Aluminium	RAL-9006	1934-60

See the homepage for »THE NIMBUS OWNERS GROUP UK«
www.nimbus.veetopia.com/Technical/Nimbus Model/Nimbus Colour Chart

PIN STRIPING

Pin striping (coach lining) is purely ornamental and is used to offer - like cosmetics - a visual impression. It does not serve any constructional purpose!

Pin striping is done manually where decorative stripes are applied with a special brush with long squirrel hair, called a pin striping brush. Nowadays, there are very few who master the skill of pin striping. Alternatively, pin striping can be done with tape either by using a coloured adhesive tape, or by using three adjacent adhesive tapes, whereas the middle one is removed to allow for 'ordinary' painting on the area between the two tapes, which are also removed afterwards.

The pinstripes of a Nimbus-C are always applied to the areas that are curved most, called 'the shoulder'.

All Nimbus-C pinstripes are in gold except for the Tivoli Blue 'Sport' version and all later (1946 – 50) blue machines. In these cases, the pin striping is in silver.

Machines, supplied to the army or post-services are never pin-striped.

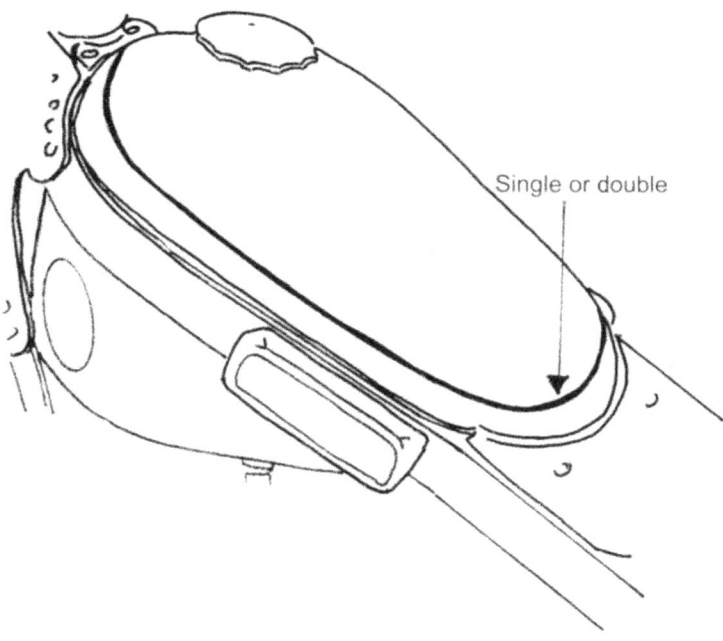

Pin striping 1301-1550
Pin striping was not applied to motorcycles from 1934-35.

Pin Striping 1501 – 7500
The 'Standard' version has no pin striping.
The versions 'Extra', 'Luxus', 'Sport' and 'Special' have a single 3 mm wide pin striping on mudguards and tank.
Wide front mudguards ('Luxus' and 'Standard Extra') have pin striping both on top (between the holes for attachment to the front fork) and at the sides (between the holes for attaching it to the telescopic fork legs).
Narrow front mudguards ('Sport' and 'Special') have pin striping on top only.
Rear mudguards have pin striping from the pivoting point at the front, under the brace (which has no pin striping) all the way back to the rear lower end.
The fuel tank has pin striping all the way around at the 'shoulder' of the curving.

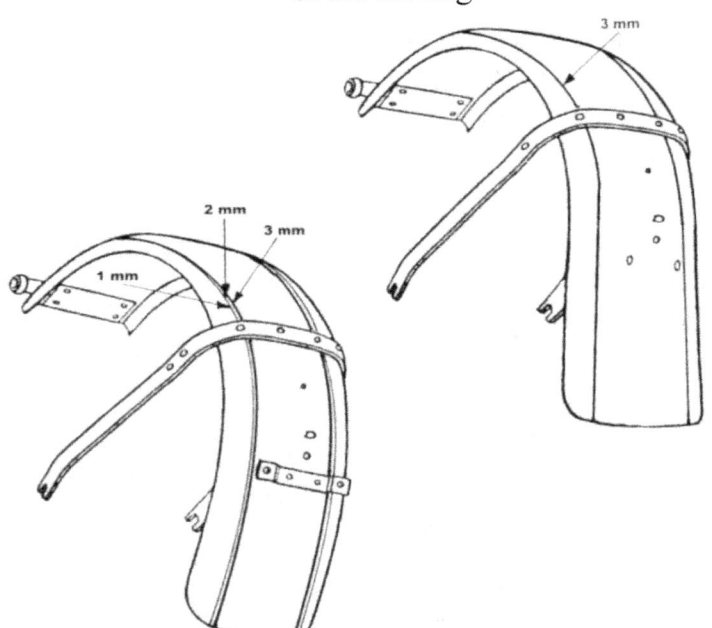

Pin striping 7501 – 14015
All versions except for machines for the army and for postal services have double pin striping on the fuel tank and on top of the mudguards. The stays of the front- and rear mudguards have no pin striping.
This double pin striping consists of a 3mm wide inner striping and a 1mm wide outer striping. The space between the stripes is about 2 mm and runs along the front mudguard over the holes for the brace.
Rear mudguards are pin-striped like 1501 – 7500, but double. The fuel tank has double pin striping all the way around at the shoulder of the curving.

NUMBERS etc.

Fabrication/production number
For all machines from 1934 up till and including March 1956, the fabrication-/production number is the same as the frame- and engine number.

The *engine* number is stamped on the left-hand side of the cylinder block. In 1934 this was done on the cylinder block flange, straight under the carburettor; the first machines had the prefix »No« added to the number.
From 1551 onward, the number was stamped on an elevated part of this flange.
For one series of machines, roughly from 2400 – 7500, the elevated part on the flange of the cylinder block with the engine number was situated somewhat to the front.

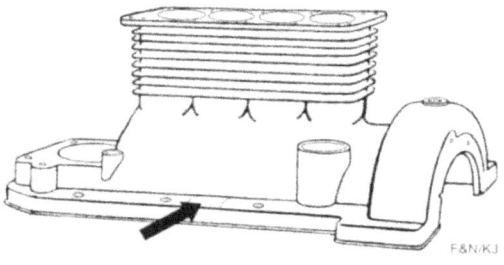

Position of the *Frame* number

1301 -2400 between the fuel tank and the seat.

- 1301 – 1550 on a make- and number (ID) plate, fitted on the clamping plate.

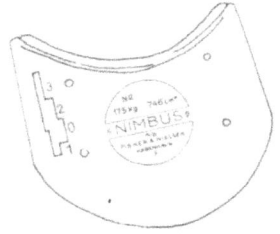

- 1551 – 2400 engraved in the clamping plate between the fuel tank and the seat.

2401 – 2550 on *two* different places:
- on the clamping plate between the fuel tank and the seat,
- and on an ID plate around the charging light on the handlebar.

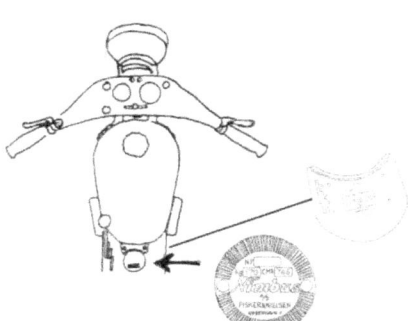

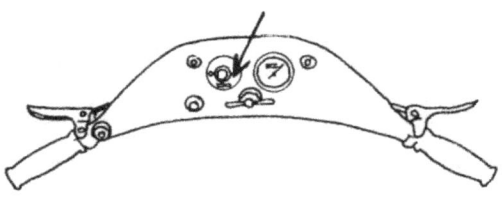

2551 – 7500 only on the ID plate around the charging light on the handlebar.
Up till around 3300, this plate showed a weight of 175 kg, thereafter 185 kg.

Some plates show the number 7, over stamped by the number 8. Normally, the plates are made of chrome-plated brass, but zinc plates have also been used.

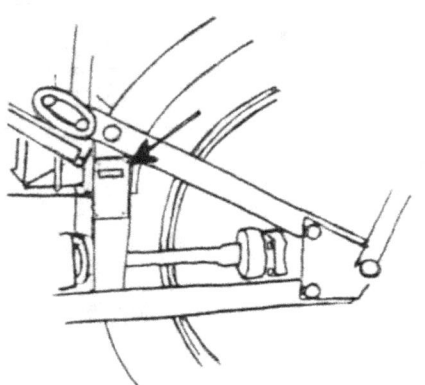

7501 – 13572 on a rectangular plate, riveted on the front-most left hand side of the rear mudguard, right below the seat. Originally, this was a brass plate, later on zinc plates were used and finally, the most common version was aluminum.

On the plates 7500 – 8000, the indicated clearance for the exhaust valves was 0.5 mm. Owners of these machines were later offered to have the plates replaced by new ones with 0.7 mm clearance. Many, but not all owners, accepted this offer. Therefore, for the numbers 7501 – 8000 both types of plates can be found.

Frame number

From engine- and production number 13573 on, this number is still on the ID plate, fitted on the foremost left hand side of the rear mudguard.

But from now on the frame has its own number, stamped on the lower, welded flange of the headstock, together with the model name »Nimbus-C«.

The first number with this identification is 15001, corresponding with production- and engine number 13573. Unfortunately, frame numbers and production numbers are from here on not sequential, but by knowing one of the two, the other one can be found in the firm's stock books. (See chapter *Stock books*)

Flywheel

The flywheel has a groove at its circumference, marked with an »I«. It is used for correctly assembling the engine and as a reference for adjusting the ignition.

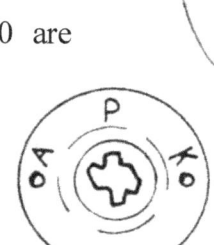

Ignition switch surround

The earliest ignition switch surrounds for 1301 – 7500 are marked 'A', 'P' and 'K'. 'A' for Off (»Afbrudt«), 'P' for Parking (»Parkering«) and 'K' for Run (»Kørsel«). The ignition key surrounds for 7501 – 14015 are unmarked.

Cylinder blocks with 'S' marking

The mark 'S' at the rim of oil filling hole in the cylinder block indicates that this engine has been assembled by the firm's employee Svend Heden during the period 1935 – ca. 1943. No other initials were found.

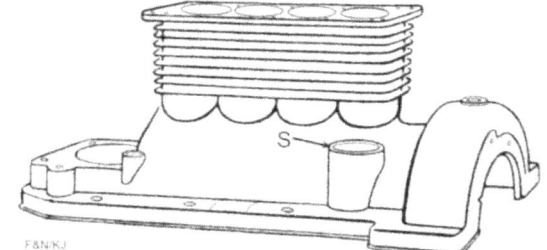

Crankshaft with 'F' marking

A crankshaft with a mark 'F' on the foremost journal face indicates that journals are hard-chrome plated.

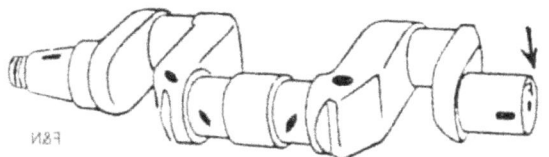

Other marks

Casting marks and -numbers on the cylinder block and -head are applied by the casting firm. There is no relationship with engine numbers.

The spare parts catalogue mentions that a triangle preceding a catalogue number means that this part is obsolete and will be replaced. The same symbol can sometimes be found on the spare parts themselves, which may have been produced – perhaps in a small series – after the part became obsolete. By the

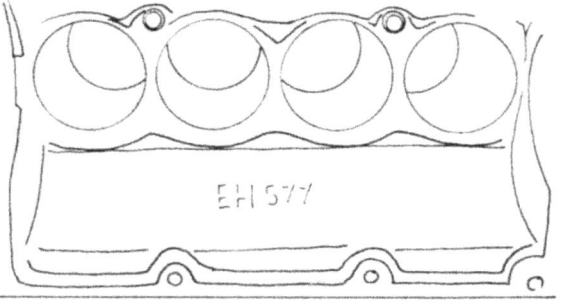

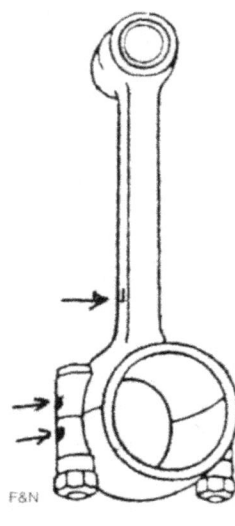

same token, square marks can be observed for parts that need to be adapted. The meaning of several types of stars is unknown. The same is true for marks like e.g. »A 50« and »G 55«.

Connecting rods

When assembling, or dissembling an engine, it is important to avoid mixing or inverting the connecting rods and bottom bearing halves (big-end caps). Therefore, the shafts of the original connecting rods are marked 1 – 2 – 3 – 4, corresponding with cylinder number 1, etc. Normally, the marks are oriented away from the lube oil pipe. Where the top- and bottom bearing, halves meet around the journal, both halves are marked with a number and/or a letter that must be adjacent to each other.

Gearwheel on camshaft

If a gearwheel with angled teeth is marked »2«, this is due to a design change as of engine no. 3000. The meaning of star-shaped marks is mostly unknown (see chapter about camshaft gear wheel).

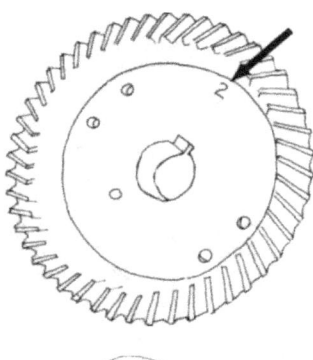

Main bottom bearing halves / cylinder block

The two bearing halves (caps) of the crankshaft's main bearings are different. To fit them correctly, they are marked '1' and '2'. The same system is used in the cylinder block for the studs of the main bearing halves.

Normally, but not in all cases, the front bearing halves are marked '1' and the rear ones '2'. There have been some wrong marks observed!

Dynamo number

The dynamo number is stamped on the edge of the dynamo end bracket, often adjacent to the cylinder block. There is no relation between the engine number and the dynamo number.

Voltage regulator number
The earliest voltage regulators, produced by A/S Fisker & Nielsen have the number stamped on the brass bottom plate. Later F&N voltage regulators were not numbered. There is no relation between engine-, dynamo- and voltage regulator numbers.

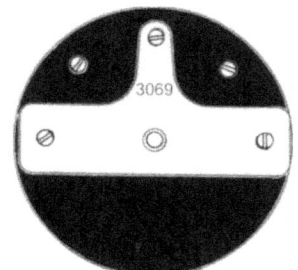

Key number
Most machines with a steering lock (after 13573) (except, but military machines), have a key, the number of which is listed in the firm's stock books. (See Chapter *Stock books*).

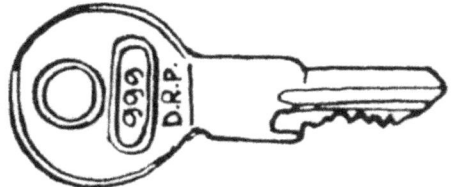

Rear Drive Ratio Identification (ID)
Machines fitted with a sidecar- or low gearing, are fitted with an angled bracket on the final drive housing identifying the installed drive ratio. This bracket is marked »LAV« (low) for early-, or »12:59« for later machines. Solo machines are not fitted with a ratio ID bracket.

Combination (ignition and lighting) switch
For easy fitting and repair of the electric wiring system, the combination switch is provided with numbers 1 – 7 and letters B – D. The wiring, the relay and other parts of the electrical system correspond with the electrical schematic.

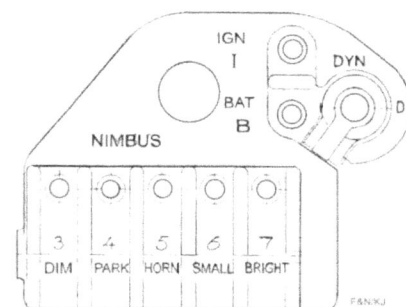

Ignition adjustment marks
The early ignition coils are marked with a casted 'S' and a 'T' on the bottom. By the same token, these characters appear on the adjustment plate on the distributor housing as from 2051 on. The 'S' and the 'T' stand for 'Sen' (late) and 'Tidlig' (early) respectively.

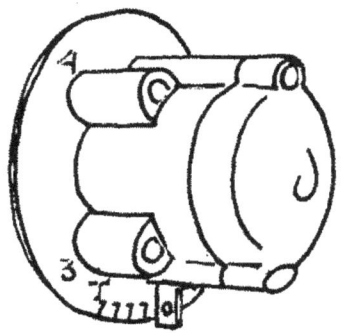

Later marking
Later Nimbus-C leaving the factory, may have had its parts marked in different ways.
In the early days, it was common use to paint- mark the ground crankshafts and the connecting rods with the following colours:
- green for 40.00 mm and green/red for 39.75 mm.
- red for 39.50 and red/white for 39.25 mm.
- white for 39.00 mm and white/yellow for 38.75 mm.
- yellow for 38.50 mm.

Defective parts
Rejected parts are marked 'DEF' and are often used in cut-open demonstration- and educational engines. These types of engines may have been dissembled and their parts erroneously used for regular engines.

DEF

Engine repair plate
Army Materials Administration, »FKF« (»Forsvarets Krigsmateriel Forvaltning«) riveted a brass plate on the flywheel housing of the cylinder block, most of the time on the left-hand side, with the following information:

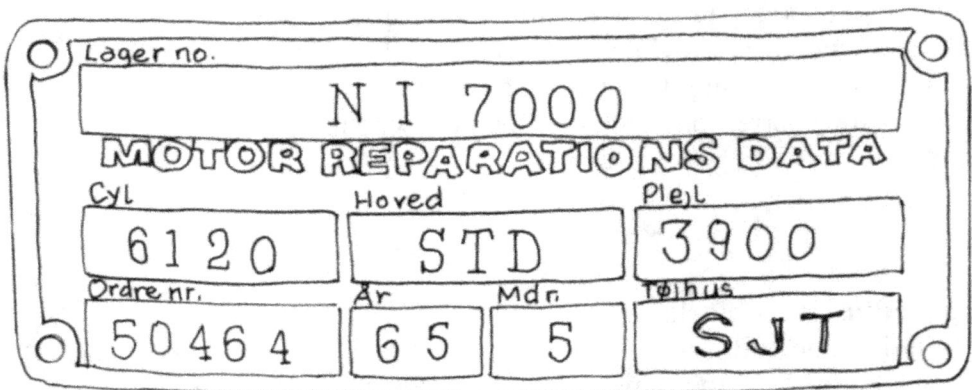

Stock No.: Normally '*NI 7000*'
Engine repair data (stamped).
'*Cyl*' with the dimension after reboring, e.g. '*Cyl 6120*' = 61.20 mm
'*Hoved*' (main bearing); may be marked '*Hoved. STD*' (standard) or '*Hoved. NYE*' (new)

'Plejl' (connecting rod) followed by four digits, e.g. *'Plejl. 3900'* = 39.00 mm.
'Certifikat nr.' (Certificate number) or *'Ordre nr'* (order number) is indicated with a trailing number and/or a letter-code.
'Aa' or *'År'* (year) is indicated with four digits, e.g. *'Aar 65'* or *'År 65'* = 1965.
'Mdr.' (month) is indicated by a number, e.g. *'Mdr. 5'* for May or *'Mdr. 12'* for December.
'Tøjhus' (armoury). There were two armouries in Denmark, which were indicated with three characters, *'Tøjhus SJT'* (Sjælland) for the Zealand armoury or *'Tøjhus JYT'* (Jylland) for the Jutlandic armoury.

NUMBER PLATES

This chapter is about Danish number plates from 1930 until 1976, the period during which Nimbus-C was produced and sold in Denmark.
The information about number plates for Nimbus-C is of course also applicable for other motorcycles, registered in Denmark.
The way of fitting the number plate can be found in the chapter about mudguards for both, the motorcycle and the sidecar.

Since 1998, it is allowed to have historically correct number plates fitted in Denmark, if the first registration of the vehicle was before April 1st, 1958. This period was later extended till March 31, 1976. This means that in Denmark a number plate can be obtained with the same appearance as a number plate from the first time the vehicle hit the road.
If the original registration number of a motorcycle can be supported by the required documentation, a plate with the corresponding number can be obtained.

Front number plates are no longer allowed at all and the use of sidecar number plates has been deleted too, so only one historically correct (rear) number plate can be obtained, whether or not the motorcycle has a sidecar.

The Danish system of obtaining a chosen number upon request applies also for motorcycles. A number plate with a chosen number has a white field with a red rim, with a combination of

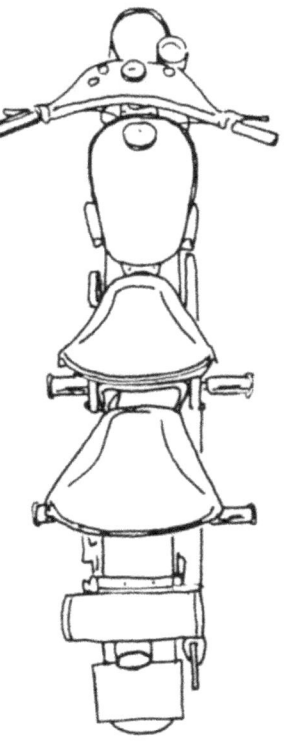

at least two letters and/or numbers with a maximum of seven, which has not been issued before. This system will no further be described here.

In 1976 white reflecting number plates with a red rim and black letters and numbers were introduced.

During the Nimbus-C production period from 1934 – 1959, number plates for non-military use were designed as follows:

Number plates issued from October 1st 1930 – June 30st 1950

Rear number plate: Flat steel plate with folded edges, black enameled with one white letter referring to the county, and under this an »Argos mark« (a hand with an eye). In addition, a sequential number of up to five white numbers. The number plate was 20 cm wide, 23 cm high and had two holes at the top and one at the bottom. This type of number plate was called 'mitten plate' or 'the one with the hand'.

Front number plate: Flat steel plate shaped to fit the front mudguard, black enameled with a white county letter, numbers and symbol at both sides, similar to the rear plate.

Sidecar number plate: Flat round steel plate with a diameter of 90 mm, black enameled with a white county letter, numbers and symbol, similar to the rear plate. It had a wide black rim and two holes for attaching it with screws, one mid top and one mid bottom. Sidecars have their own sequential numbers, not related to the number of the motorcycle.

Number plates issued from July 1st 1950 – March 31, 1953.

Rear number plate: Flat steel plate with folded edges and either black enameled with a white a letter and numbers or (no VAT) yellow enameling with a black letter and numbers. The county number has the same size as the (up to) five-digit sequential number. The plate is 20 cm wide, 23 cm high and has two holes' mid top and mid bottom.
Note: In Denmark, industrial products are divided into three VAT

categories, full tax, half tax or no tax to be paid.

Front number plate: Flat steel plate, shaped to fit the front mud guard, either black enameled with a white letter and numbers or (no VAT) yellow enameling with a black letter and numbers on both sides, similar to the rear plate.

Sidecar number plate: Flat round steel plate with a diameter of 90 mm, either black enameled with white numbers and rim (full VAT) or yellow enameled with black numbers and rim (no VAT). Sidecars have no county letter, but the number of the police district instead, above a horizontal stripe. The sidecar sequential number is placed below the stripe.

Number plates issued from April 1st 1953 – March 31, 1958

Rear number plate:
Full VAT: like rear number plate for 1950 – 1953.
Half VAT: the sequential number is white on a black background. The county letter is black on a yellow background, also called 'parrot eye'.
No VAT: Similar to the yellow plate from 1950 – 1953.

Front number plate:
Full VAT: Similar to the front number plate from 1950 – 1953.
Half VAT: The sequential number is white in a black field and the county letter is black in a yellow field.

No VAT: Similar to the yellow front number plate from 1950 – 1953. The front number plate was forbidden as from April 1st 1956. Therefore, a front number plate with "Parrot eye" is rare.

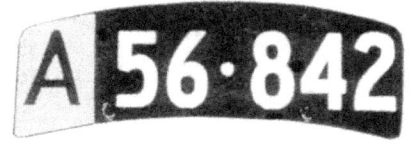

Sidecar number plate: Similar to sidecar plates from 1950 – 1953. For half VAT with 'parrot eye' number plates, it was mandatory to have a sidecar fitted for transportation or goods. Therefore, the sidecar number plate was yellow with a black letter, numbers and rim.

TWO LETTER NUMBER PLATES

Number plates issued from April 1st, 1958 – March 31, 1976

Rear number plate: Flat steel plate with folded edges, black enameled and white letters and numbers. The first letter represents the county and the second one the police district. The letters and numbers have the same size. The white five-digit sequential number indicates the vehicles use: 10,000 – 16,999: private use, 17,000 – 17,999: rental- and instruction motorcycles. The sequential numbers 18,000 – 19,999 were applied to motorcycles for transportation of goods.

If the vehicle was registered under 'no VAT', it had yellow enameled plates with black letters and numbers.

If the vehicle was registered under 'half VAT', it bore black enameled plates with white numbers and black letters in a yellow field. The plate was 23 cm wide and 20 cm high with vertical brackets for dedicated mounting bolts.

Front number plate: N/A.

Sidecar number plates: Flat round steel plate with a diameter of 90 mm, either black enameled with white letters, numbers and rim or yellow enameled with black letters, numbers and rim. There were two fitting holes, one in the middle of the top side, and one in the middle of the bottom side. Sidecars bore the same county- and police district numbers as the motorcycle but they had their own sequential number, from 50 – 2999.

MILITARY NUMBER PLATES

Military number plates for the period 1934 – 59:
Number plates for Ground Forces:
- Until August 29, 1943 and 1945 – 46: Black painted steel plate with affixed brass Ground Force Mark (leaf garland crest wreath with a crown) and a white, five-digit number.

- After 1946: A red enameled steel plate with the Ground Force Mark in black and yellow with white numbers.
- After 1976: Black steel plate with a glued-on black Ground Force Mark in black and yellow and similar white numbers.

Nimbus-C motorcycles in mobilization depots may be fitted with 1976-type of number plate.

Number plates for the Navy:
- Until August 29, 1943 and from 1945 – 46: Black painted steel plates with a crown and a white, one to five-digit number.
- After 1946: Navy blue enameled steel plate with a yellow anchor and white numbers.

Number plates for the air troops:
- Until and including 1950:
- - Army air Force troops: Army number plates
- - Navy air troops: Navy number plates
- After 1950, Danish Air Force:
- - Light blue enameled steel plate with yellow wings and white numbers.

Number plates for Civil Protection and Civil Defense:
- Until and including 1949: Civil number plates: Black enamelled with one white letter, the Argos symbol and white numbers.
- After 1949: Blue/gray enameled steel plate with CF (Civil Forsvar = Civil Defense) and a yellow crown with white numbers.

OVERVIEW

The next pages describe the Nimbus-C production over the period 1934 – 59, first one year at a time and further on a couple of years together. The purpose of the text and the accompanying drawings is to provide an overview of the technical development. In addition, sales prices for new Nimbus-C motorcycles, as delivered by the factory, are provided. To be able to compare the prices, both, the factory- and dealer prices, are given. The dealer price includes delivery- and registration costs and VAT, but not road tax and insurance.

The overview is not exhaustive, but covers just the broad lines. The details need to be filled in with the other information that can be found elsewhere in the book. The paragraph *Colours* e.g. describes only the color of the enamel of the handlebars, the top of the fuel tank and the mudguards. The enamel or surface treatment of the other parts can be found in the chapter *Identification*, pararaph *Versions*.

1934

Year:	1934 (1301 - 1500; 200 in total)
	(1501 – 1550, assembled in 1935, are considered as 1934; 50 in total)
Version:	'Standard'
Color:	Black, no pin striping (a very few were supplied in non-black)
Price:	Factory price: DKK 1,880, dealer price: appr. DKK 2,070, including pillion seat.
Number:	Frame- and engine number are identical. Make- and frame number are on a round brass plate, riveted on the clamping plate between fuel tank and seat.
Engine:	18 HP, Compression 1:5, pressure-lubricated crankshaft, splash-lubricated camshaft.
Carburettor:	Nimbus '34-1. Matt nickel plated fuel pipe without pigtail.
	Rotating fuel tap Zöblitz AG (MZAG), later a double slide valve of same make.
Dynamo:	Dynamo '34: third-brush dynamo 6V, 25 – 30 W with oil pressure operated cut-out switching.
Gearbox:	Early version (see Andersen below) constructed for the rigid (uncushioned) drive shaft. Hand operated.
Transmission:	Foot- and hand operated clutch, rigid drive shaft, ratio 14 : 56.
Fuel tank:	Outlet at the front left side of the base.
Frame:	Single head tube with welded reinforcement at the bottom. Small hole in the baffle plate for rigid driveshaft. One hole in the left hand fish plate of the frame to attach the final drive housing.
Front fork:	Front fork '34: telescopic with grease lubrication. Leather sleeves. No keyway in the handlebars. Friction damper with cross pin handgrip.

Handlebars:	Flat, with visible rivets. Round headed screws for combination switch. 19 mm hex Centre bolt for steering damper.
Front mudguard:	Front arch follows the wheel. Deep side valances.
Rear mudguard:	Rounded at the rear end from the tail light down. Reduced diameter of the enlarged ends of the seat hinge pivot support.
Centre stand:	Without reinforcement around the hinges and the feet.
Seat:	Compression coil springs with eyes.
Pillion seat:	Hinges normally within the frame. Compression coil springs with eyes.
Wheels:	Front wheel with straight spokes. Rear wheel with narrow hub.
Brakes:	Front and rear brakes 150 mm. From 1526 the construction of front brake was reinforced.
Speedometer:	'VDO' make with external, indirect lighting. Ratio 1:3 drive.
Lights:	Head light 'Riemann' make, with arched front glass. Tail light with a bulge and a cast partition. 6 mm holes for the wires at both sides. Matt red celluloid plate with text »STOP«.
Horn:	'Riemann' with reinforced back.

Note: The first 250 units from 1934 – 35 were a kind of prototype. During this period, several changes were introduced. A couple of crucial changes were implemented as from 1551. (See Andersen (1966): http://www.geutskens.eu)

1935

Year:	1935. The numbers 1501 - 1550 are considered to be from 1934, 1551 – 2014 are 'really' from 1935, total 514 units.
Version:	'Standard' and 'Luxus'
Colours:	'Standard': black without pin striping; 'Luxus': black, red or green with single pin striping.
Price:	Factory: 'Standard' DKK 1,880, 'Luxus' DKK 1,980. Dealer price: appr. DKK 2,070, 'Luxus' appr. DKK 2,180 including pillion seat.
Number:	Frame- and engine number are identical. Make-, frame number and weight (175 kg) are engraved in the clamping plate between seat and the fuel tank.
Engine:	18 HP, Compression 1:5, splash-lubricated crankshaft, dip-lubricated camshaft
Carburettor:	Nimbus '34-1. Matt nickel plated fuel pipe with pigtail. Dual German slide fuel valve Zöblitz AG (MZAG).
Dynamo:	Till 1550: dynamo '34: third-brush dynamo 6V, 25 – 30 W with oil pressure operated cut-out switching. From 1551: two brush dynamo 6/8 V, 70 W and a voltage regulator under seat.
Gearbox:	Early version, constructed for the rigid (un-cushioned) drive shaft. Hand operated.
Transmission:	Foot- and hand operated clutch, rigid drive shaft, ratio 14 : 56. On special demand: ratio 12:59.
Fuel tank:	Outlet at the front left side of the base.
Frame:	Double head tube. Small hole in the baffle plate for rigid driveshaft. One hole in the left hand fish plate of the frame to attach the final drive housing.
Front fork:	Front fork '34: telescopic with grease lubrication. Leather sleeves. No keyway in the handlebars. Friction damper with cross pin handgrip.
Handlebars:	Flat, rivets not visible. Round headed screws for combination switch. 19 mm hex Centre bolt for steering damper.
Front mudguard:	Front arch follows the wheel. From number plate, straight forward. Deep side valances.
Rear mudguard:	Somewhat round top. Rounded at the rear end from the tail light down.
Centre stand:	Without reinforcement around the hinges. Welded-on feet.
Seat:	Till 1900 compression coil springs with eyes. From 1901 compression coil springs with angled plates.
Pillion seat:	Hinges outside the frame. Till 1900 compression coil springs with eyes. From 1901 compression coil springs with angled plates.

Wheels:	Front wheel with straight spokes.
	Rear wheel with narrow hub.
Brakes:	Front and rear brakes 150 mm.
Speedometer:	'VDO' make with external, indirect lighting. Ratio 1:3 drive.
Lights:	Head light 'Riemann' make, with arched front glass.
	Tail light with a bulge and a cast partition. Matt red celluloid plate with text »STOP«.
Horn:	'Riemann'.

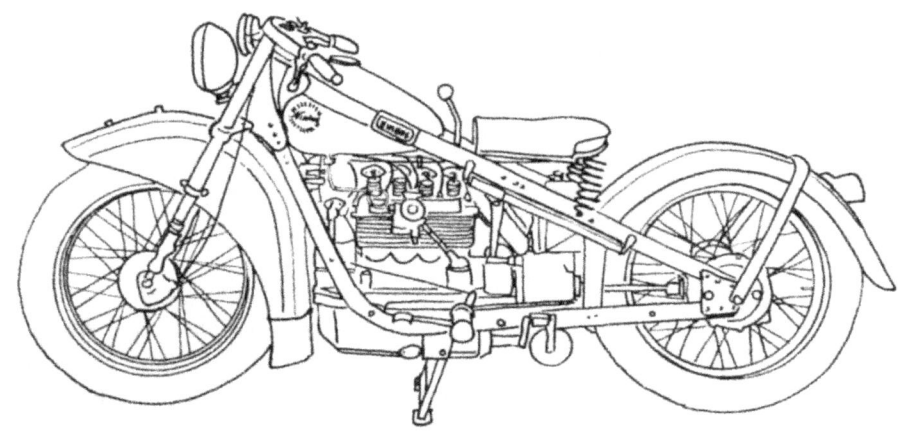

1936

Year:	1936, 2015 – 2646 in total 632 units.
Versions:	'Standard', 'Standard Extra' and 'Luxus'
Colours:	'Standard': black without pin striping.
	'Standard Extra' and 'Luxus': red or green with single pin striping.
Prices:	Factory: 'Standard': January 1936: DKK 1,925, October: DKK 1,965.
	'Standard Extra': DKK 1,975; 'Luxus': DKK 2.025 with pillion seat.
	Dealer price October: 'Standard' appr. DKK 2,240, 'Standard-Extra' appr. DKK 2,250, 'Luxus' appr. DKK 2,310 including pillion seat.
Number:	Frame- and engine number are identical for 2015 – 2400: Make and frame number are engraved in the clamping plate between seat and the fuel tank.
	2401 – 2550: Make- frame number and weight (175 kg) on the clamping plate between the seat and the fuel tank as well as on a plate around the charging light on the handlebars.
	From 2551: Make- frame number and weight (175 kg) only on a plate around the charging light on the handlebars.

Engine:	18 HP, Compression 1:5, splash-lubricated crankshaft, dip-lubricated cam-shaft.
Carburettor:	2015 – 2515: Nimbus '34-1. From 2516: Nimbus '34-2. Matt nickel plated fuel pipe with pigtail. Dual German slide fuel valve.
Dynamo:	Dynamo '35: two brush dynamo 6/8 V, 70 W. Till 2050: voltage regulator '35. From 2051: voltage regulator '36 under seat.
Gearbox:	Early version. Till 2550: constructed for the rigid (un-cushioned) drive shaft. From 2551: constructed for the cushioned drive shaft. Hand operated.
Transmission:	Foot- and hand operated clutch. Till 2550: rigid drive shaft. From 2551: cushioned drive shaft. 'Standard': ratio 12:59. 'Standard-Extra' and 'Luxus': Ratio 14:56.
Fuel tank:	Till 2050 outlet at the front left side of the base.
Frame:	Double head tube. Till 2560 with tiny hole in the baffle plate for rigid driveshaft. From 2561 with larger hole in the baffle plate for cushioned driveshaft. Till 2560 with one hole in the left hand fish plate of the frame to attach the final drive housing. From 2561 with two holes for attaching the final drive housing.
Front fork:	Front fork '34: telescopic with grease lubrication. Leather sleeves. Till 2550 without keyway in the handlebars. From 2551 front fork '36 with keyway in the handlebars for a safety washer. Till 2600 friction damper with cross pin handgrip. From 2601 friction damper with Bakelite handgrip.
Handlebars:	Flat, rivets not visible. Round headed screws for combination switch. Till 2600 19 mm hex centre bolt for steering damper. From 2601 28 mm hex centre bolt.
Front mudguard:	Front arch follows the wheel. From number plate, straight forward. Deep side valances.
Rear mudguard:	Somewhat rounded top. Rounded at the rear end from the tail light down.
Centre stand:	Without reinforcement around the hinges. Welded-on feet.
Seat:	Compression coil springs with angled plates.
Pillion seat:	Hinges outside the frame. Compression coil springs with angled plates. Handgrip made from 14 mm chrome plated round steel or brass.
Wheels:	Front wheel with straight spokes. Rear wheel with narrow hub.

Brakes:	Front and rear brakes 150 mm.
Speedometer:	Till 2550 'VDO' make with external, indirect lighting. Ratio 1:3 drive.
	From 2551 'VDO' with built-in lighting, ratio 1:3.
Lights:	Head light 'Riemann' make, with arched front glass.
	Till 2050 tail light with a bulge and a cast partition. Matt red celluloid plate with text »STOP«.
	From 2051 polished tail light with a partition with loose wall and a clear red/yellow celluloid plate.
Horn:	'Riemann'.

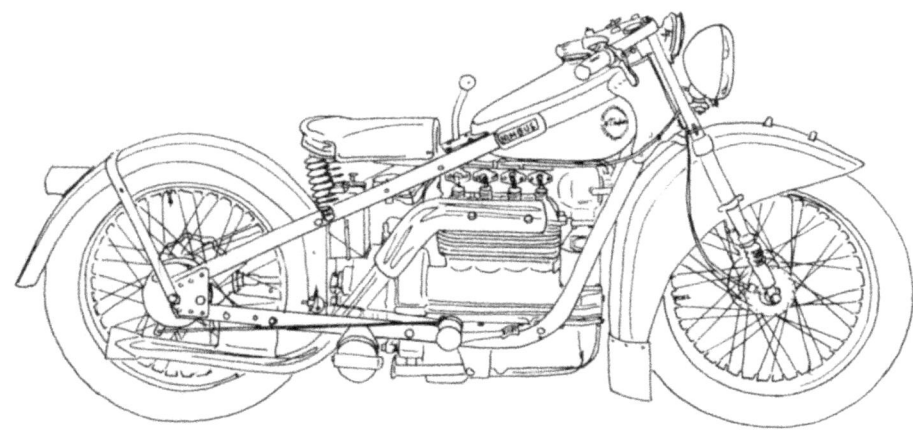

1937

Year:	1937, 2647 – 3489 in total 843 units.
Versions:	'Standard'', 'Luxus' and 'Sport' (first 'Sport' no. 2860)
Colours:	'Standard': black without pin striping.
	'Luxus': red or green with single gold pin striping.
	'Sport': blue with single silver pin striping.
Prices:	Factory: 'Standard' DKK 1,975, 'Luxus' DKK 2,035, 'Sport' DKK 2,135
	Dealer price: 'Standard' appr. DKK 2,230, 'Luxus' appr. DKK 2,305, 'Sport' appr. DKK 2,420 including pillion seat.
Number:	Frame- and engine numbers are identical.
	Till 3000: Make- frame number and weight (175 kg) on a plate on the handlebars.
	From appr. 3000: Indicated weight 185 kg.
Engine:	'Standard': 18 HP, compression 1:5.
	'Sport': 22 HP, compression 1:5,7.
	Splash-lubricated crankshaft, dip-lubricated camshaft.

Carburettor:	Nimbus '34-2. Matt nickel plated fuel pipe with pigtail.
	Some 'Sport' versions (3001 – 3050) with carburettor Nimbus '38 and fuel hose.
	Dual German fuel slide valve.
Dynamo:	Dynamo '35: two brush dynamo 6/8 V, 70 W.
	Voltage regulator '36 under seat.
Gearbox:	Early version. Constructed for the cushioned drive shaft.
Transmission:	'Standard' with hand operated gear shift and foot- and hand operated clutch.
	'Sport' with foot operated gear shift and hand operated clutch.
	Cush drive shaft.
	'Standard': ratio 12:59.
	'Luxus' and 'Sport': Ratio 14:56.
Fuel tank:	Outlet at the mid left side of the base.
	Dedicated transfer for 'Sport' version.
Frame:	Double head tube.
	Hole in the baffle plate for cushioned driveshaft.
	Two holes in the left hand fish plate of the frame for attaching the final drive housing.
	Till 2900 centre stand plates without reinforcement.
	From 2901 centre stand plates with reinforcement.
Front fork:	Till 3200 front fork '36, telescopic sprung with grease lubrication.
	From 3201(including some 'Sport' version units, 3001 – 3050) front fork '37/38, telescopic sprung with grease lubrication and damping.
	Leather sleeves. Keyway in the handlebars for a safety washer.
Handlebars:	Flat, rivets not visible. Round headed screws for combination switch.
	28 mm hex centre bolt.
	Friction damper with Bakelite handgrip.
Front mudguard:	Front arch follows the wheel. From number plate straight forward.
	Deep side valances for 'Standard' and 'Luxus'.
	Shallow valances for 'Sport'.
Rear mudguard:	Somewhat rounded top. Rounded at the rear end from the tail light down.
Centre stand:	Till 2900 without reinforcement around the hinges and the feet.
	From 2901 with reinforcement around the hinges and the feet.
	Welded-on feet.
Seat:	Compression coil springs with angled plates.
Pillion seat:	Hinges outside the frame.
	Compression coil springs with angled plates.
	Handgrip made from 14 mm chrome plated round steel or brass.
Wheels:	Front wheel with straight spokes.
	Rear wheel with narrow hub.
Brakes:	Front brakes 150 mm. Rear brakes till 3000 and 3150 – 3196 150 mm.
	From 3001 – 3150 and from 3197 180 mm with pressed brake drums.
Speedometer:	'VDO' with built-in lighting, ratio 1:2.

Lights:	Head light till 3000 "Riemann" make, with arched front glass. From 3001 with flat, corrugated front glass and double reflector. Polished tail light with a partition with loose and a clear red/yellow celluloid plate.
Horn:	'Riemann'.

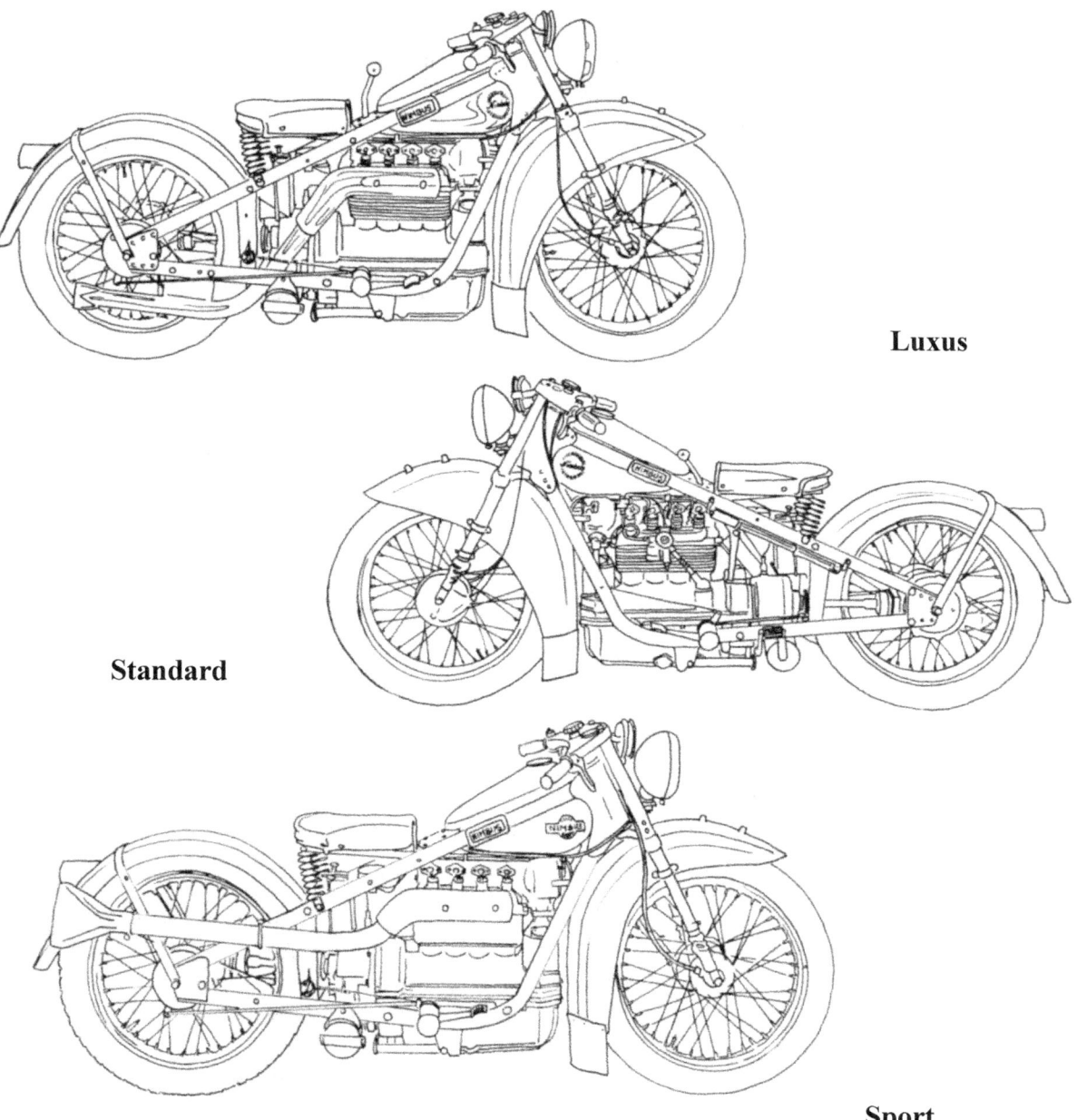

Luxus

Standard

Sport

1938

Year:	1938, 3490 – 4426, in total 9373 units.
Versions:	'Standard', 'Luxus' and 'Sport'
Colours:	'Standard': black without pin striping. 'Luxus': red or green with single gold pin striping. 'Sport': blue with single silver pin striping.
Prices:	Factory: 'Standard' DKK 2,045, 'Luxus' DKK 2,105, 'Sport' DKK 2,205 including pillion seat. Dealer price: 'Standard'. DKK 2,045, 'Luxus' DKK 2,305, 'Sport' DKK 2,205 including pillion seat. Extra charge for foot operated gear shift on 'Standard' and 'Luxus' DKK 30.-
Number:	Frame- and engine numbers are identical. Make- frame number and weight (185 kg) on a plate on the handlebars.
Engine:	'Standard': 18 HP, compression 1:5. 'Sport': 22 HP, compression 1:5,7. Splash-lubricated crankshaft, dip-lubricated camshaft.
Carburettor:	Till 3863 Nimbus '34-2. Matt nickel plated fuel pipe with pigtail. From 3864 with carburettor Nimbus '38 and fuel hose. Dual German fuel slide valve.
Dynamo:	Dynamo '35: two brush dynamo 6/8 V, 70 W. Voltage regulator '36 under seat.
Gearbox:	Early version. Constructed for the cushioned drive shaft.
Transmission:	'Standard' with hand or foot operated gear shift and foot- and hand operated clutch. 'Sport' with foot operated gear shift and hand operated clutch. 'Standard': ratio 12:59. 'Luxus' and 'Sport': Ratio 14:56.
Fuel tank:	Outlet at the mid left side of the base. Dedicated transfer for 'Sport' version.
Frame:	Double head tube. Hole in the baffle plate for cushioned driveshaft. Two holes in the left hand fish plate of the frame for attaching the final drive housing. Centre stand plates with reinforcement.
Front fork:	Front fork '37/38, telescopic sprung with grease lubrication and damping. Leather sleeves. Keyway in the handlebars for a safety washer. Friction damper with Bakelite handgrip
Handlebars:	Flat, rivets not visible. Round headed screws for combination switch. 28 mm hex centre bolt.
Front mudguard:	Front arch follows the wheel. From number plate straight forward. Deep side valances for 'Standard' and 'Luxus'. Shallow valances for 'Sport'.

Rear mudguard:	Somewhat rounded top. Rounded at the rear end from the tail light down.
Centre stand:	Reinforcement around the hinges. Welded-on feet
Seat:	Compression coil springs with angled plates.
Pillion seat:	Hinges outside the frame. Compression coil springs with angled plates. Handgrip made from 14 mm chrome plated round steel or brass.
Wheels:	Front wheel with straight spokes. Rear wheel with narrow or wide hub.
Brakes:	Front brakes 150 mm. Rear brakes 180 mm with pressed brake drums.
Speedometer:	'VDO' with built-in lighting, ratio 1:2.
Lights:	Head light 'Riemann' make, with flat, corrugated front glass and double reflector. Polished tail light with a loose partition and a clear red/yellow celluloid plate.
Horn:	'Riemann'.

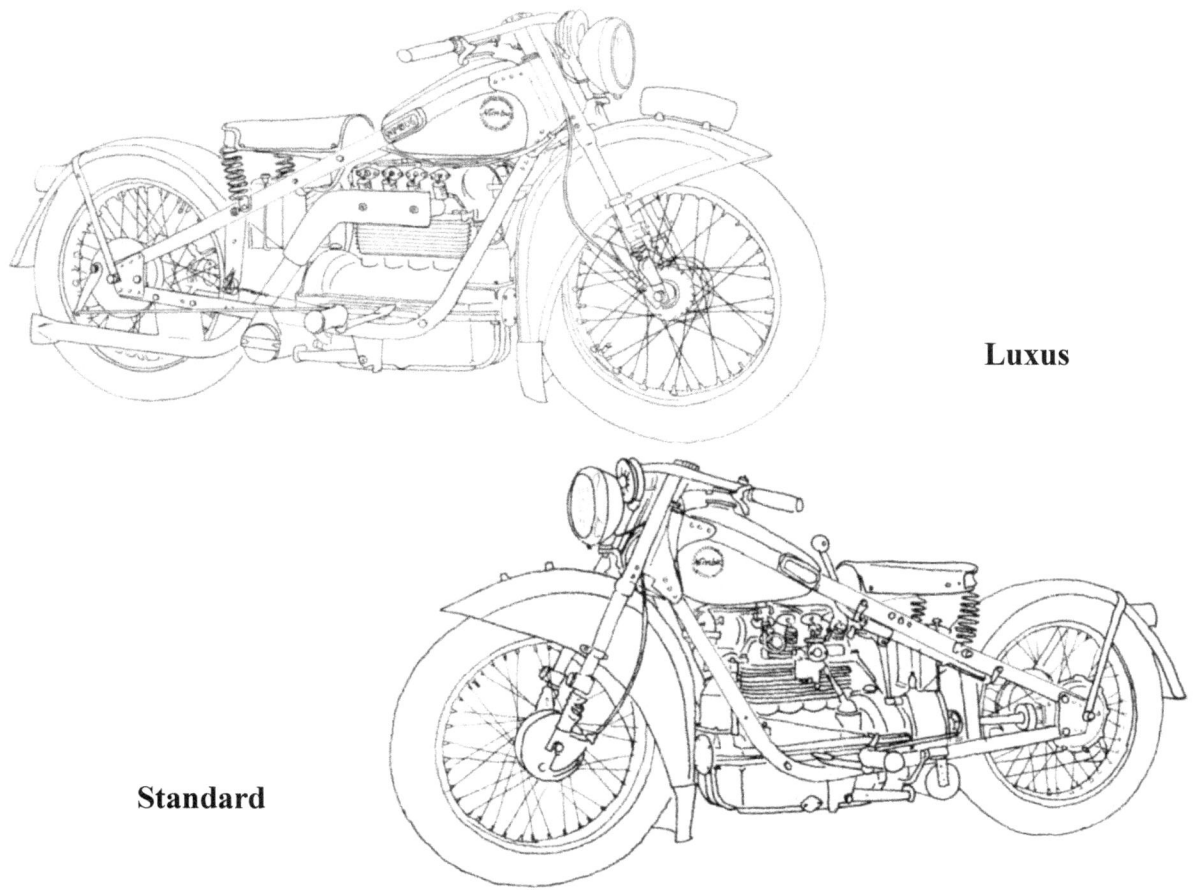

Luxus

Standard

1938 Sport

1939

Year:	1939, 4427 – 5512, in total 1086 units.
Versions:	'Standard', 'Luxus', 'Special' and 'Sport'
Colours:	'Standard': black without pin striping. 'Luxus': black, red or green with single gold pin striping. 'Special': grey (polychromatic grey) or ivory yellow. Grey with single gold pin striping. 'Sport': blue with single silver pin striping.
Prices:	Factory: 'Standard' DKK 1,965, 'Luxus' DKK 2,105, 'Special' and 'Sport' DKK 2,205 including pillion seat. Dealer price: 'Standard'. DKK 2,412.50, 'Luxus' DKK 2,517.50, 'Special' and 'Sport' DKK 2,557.50 including pillion seat. Extra charge for foot operated gear shift on 'Standard' and 'Luxus' DKK 35.-
Number:	Frame- and engine numbers are identical. Make- frame number and weight (185 kg) on a plate on the handlebars.
Engine:	'Standard': 18 HP, compression 1:5. 'Sport' and 'Special': 22 HP, compression 1:5,7. Splash-lubricated crankshaft, dip-lubricated camshaft.
Carburettor:	Nimbus '38 with fuel hose. Till appr. 4500: Dual German fuel slide valve. From appr. 4500: German rotating fuel valve.
Dynamo:	Till appr. 5000: Dynamo '35. From appr. 5000: Dynamo '39, two brush dynamo 6/8 V, 70 W. Voltage regulator '36 under seat.
Gearbox:	Early version. Constructed for the cushioned drive shaft.

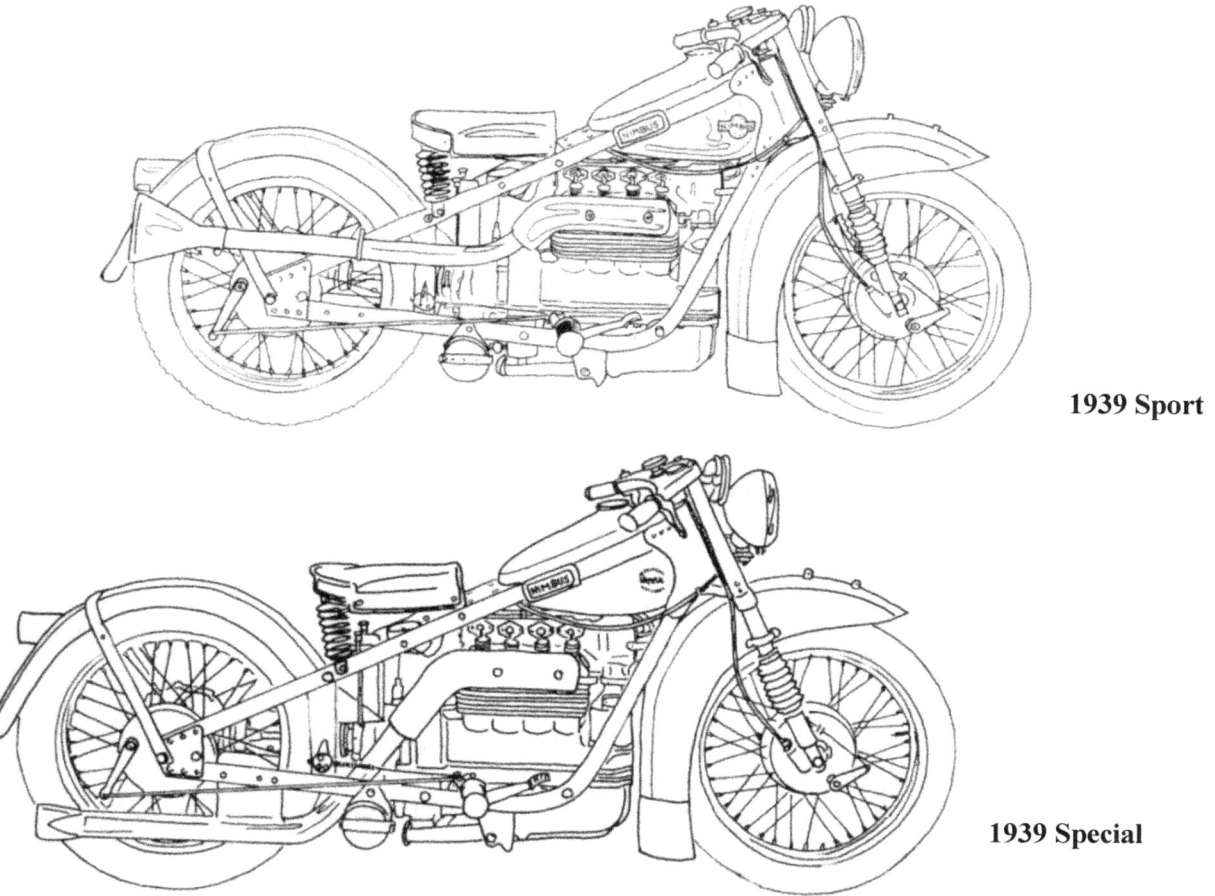

1939 Sport

1939 Special

Transmission:	'Standard' with hand or foot operated gear shift and foot- and hand operated clutch.

'Sport' and 'Special' with foot operated gear shift and hand operated clutch.
Cush drive shaft.
'Standard': ratio 12:59.
'Luxus': ratio 12:59 or 14:56.
'Sport' and 'Special': ratio 14:59

Fuel tank: Outlet at the mid left side of the base. Dedicated transfer for 'Sport' and 'Special' version.

Frame: Double head tube.
Hole in the baffle plate for cushioned driveshaft.
Two holes in the left hand fish plate of the frame for attaching the final drive housing.
Centre stand plates with reinforcement.

Front fork:	Till 4666 and 5000 – 5100 front fork '37/38, telescopic sprung with grease lubrication and damping. Some numbers prior to 4667, 4667 – 4900 and from 5101 front fork '39. Leather sleeves. Friction damper with Bakelite handgrip.
Handlebars:	Flat, rivets not visible. Round headed screws for combination switch. 28 mm hex centre bolt.
Front mudguard:	Front arch follows the wheel. From number plate straight forward. Deep side valances for 'Standard' and 'Luxus'. Shallow valances for 'Sport' and 'Special'.
Rear mudguard:	Somewhat rounded top. Rounded at the rear end from the tail light down.
Centre stand:	Reinforcement around the hinges. Welded-on feet.
Seat:	Compression coil springs with angled plates.
Pillion seat:	Hinges outside the frame. Compression coil springs with angled plates. Handgrip made from 14 mm chrome plated round steel or brass.
Wheels:	Front wheel with straight spokes. Rear wheel with narrow or wide hub.
Brakes:	For fore fork '37/38 front brake 150 mm. For fore fork '39 front brake 180 mm. Rear brakes 180 mm with pressed brake drums.
Speedometer:	'VDO' with built-in lighting, ratio 1:2.
Lights:	Head light 'Riemann' make, with flat, corrugated front glass and double reflector. Polished tail light with a loose partition and a clear red/yellow celluloid plate.
Horn:	'Riemann'.

1940 – 46

Year:	1940 – 46 5513 - 7064, in total 1552 units.
Versions:	'Standard', 'Luxus', 'Special' and 'Sport'
Colours:	'Standard': black. 'Luxus': black, red or green. 'Special': grey (lavender grey or polychromatic grey) or yellow. 'Sport': blue.
Prices:	Factory price **1940:** 'Standard' DKK 2,225, 'Luxus' DKK 2,295, 'Special' and 'Sport' DKK 2,315. Dealer price 1940: 'Standard'. DKK 2,595, 'Luxus' DKK 2,675, 'Special' and 'Sport' DKK 2,700 including pillion seat. Extra charge for foot operated gear shift on 'Standard' DKK 35.-. Factory price **1943:** 'Standard' DKK 2,670, 'Luxus' DKK 2,754, 'Special'

	and 'Sport' DKK 2,778.
	Dealer price 1943: 'Standard'. DKK 3,281.90, 'Luxus' DKK 3,389.90, 'Special' and 'Sport' DKK 3,420.70 including pillion seat.
	Extra charge for foot operated gear shift on 'Standard' DKK 46.30.
	Factory price **1946** (for those machines that were available): appr. DKK 3,675.
	Dealer price 1946 (if possible): appr. DKK 4,580 including pillion seat.
Number:	Frame- and engine numbers are identical.
	Make- frame number and weight (185 kg) on a plate on the handlebars.
Engine:	'Standard': 18 HP, compression 1:5. 'Sport' and 'Special': 22 HP, compression 1:5,7. Splash-lubricated crankshaft, dip-lubricated camshaft.
Carburettor:	Nimbus '38 with fuel hose. German rotating fuel valve.
Dynamo:	Dynamo '39, two brush dynamo 6/8 V, 70 W. Voltage regulator '36 under seat.
Gearbox:	Early version. Constructed for the cushioned drive shaft.
	'Standard' with hand operated gear shift and foot- and hand operated clutch.
	'Sport' and 'Special' with foot operated gear shift and hand operated clutch.
Transmission:	'Standard': ratio 12:59.
	'Luxus': ratio 12:59 or 14:56.
	'Sport' and 'Special': ratio 14:59
Fuel tank:	Outlet at the mid left side of the base. Dedicated transfer for 'Sport' and 'Special' version.
Frame:	Double head tube.
	Hole in the baffle plate for cushioned driveshaft.
	Two holes in the left-hand fish plate of the frame for attaching the final drive housing.
	Centre stand plates with reinforcement.
Front fork:	Front fork '39. Rubber sleeves. Friction damper with Bakelite handgrip.
Handlebars:	Flat, rivets not visible. Round headed screws for combination switch.
	28 mm hex centre bolt.
Front mudguard:	Front arch follows the wheel. From number plate straight forward.
	Till about 6000: Deep side valances for 'Standard' and 'Luxus'.
	Shallow valances for 'Sport' and 'Special'.
	From about 6000: Shallow valances for all versions.
Rear mudguard:	Somewhat rounded top. Rounded at the rear end from the tail light down.
Centre stand:	Reinforcement around the hinges.
	Welded-on feet
Seat:	Compression coil springs with angled plates.
Pillion seat:	Hinges outside the frame. Compression coil springs with angled plates.

	Handgrip made from 14 mm chrome plated round steel or brass.
Wheels:	Front wheel with straight spokes. Rear wheel with wide hub.
Brakes:	Front brake 180 mm. Rear brake 180 mm with pressed brake drums.
Speedometer:	'VDO' with built-in lighting, ratio 1:2.
Lights:	Head light 'Riemann' make, with flat, corrugated front glass and double reflector.
	Polished tail light with a loose partition and a clear red/yellow celluloid plate.
Horn:	'Riemann'.
Note:	The machines within the range 5513 – 7064 were not manufactured or distributed sequentially.
	In 1944 – 46 a couple of machines were supplied without battery and inner tubes or tyres. Some standard colours were no longer available. Accessories varied.

1947

Year:	1947, 7065 - 7500, in total 436 units.
Versions:	'Special'. All machines from 1947 – 53 were called 'Special'.
Colour:	Black with single gold pin striping
Prices:	Factory price DKK 3,675. Dealer price DKK 4,583. including pillion seat.
Number:	Frame- and engine numbers are identical. Till 7246 make- frame number and weight (185 kg) on a plate on the handlebars. From 7247 make- and frame number (see chapter "Numbers etc.") on a plate at the left-hand side below the seat.
Engine:	22 HP, compression 1: 5,7. Splash-lubricated crankshaft, dip-lubricated camshaft.
Carburettor:	Nimbus '38 with fuel hose. German rotating fuel valve.
Dynamo:	Dynamo '39, two brush dynamo 6/8 V, 70 W. Voltage regulator '36 under seat.
Gearbox:	Transition model, which means a new model, with the 3rd gear having straight cut teeth. Hand operated gear shift and foot and hand operated clutch or foot operated gear shift with hand operated clutch.
Transmission:	Cushioned drive shaft. Ratio 12:59 or 14:56.
Fuel tank:	Outlet at the mid left side of the base. Dedicated transfer for version 'Special'.
Frame:	Double head tube. Hole in the baffle plate for cushioned driveshaft. Two holes in the left-hand fish plate of the frame for attaching the final drive housing. Centre stand plates with reinforcement.
Front fork:	Front fork '39 or the later transition model, partly like front fork '48. Rubber sleeves. Friction damper with Bakelite handgrip.
Handlebars:	Till 7246: 5° bent upward, rivets not visible. From 7247: 10° bent upward with no hole for the speedometer. Ring-shaped speedometer holder welded at underside. Round headed screws for combination switch. 28 mm hex centre bolt.
Front mudguard:	Front arch follows the wheel. From number plate straight forward. Shallow valances.
Rear mudguard:	Somewhat rounded top. Rounded at the rear end from the tail light down.

Centre stand:	Reinforcement around the hinges. Welded-on feet
Seat:	Compression coil springs with angled plates.
Pillion seat:	Hinges outside the frame. Compression coil springs with angled plates. Handgrip made from 14 mm chrome plated round steel or brass.
Wheels:	Front wheel with straight spokes. Rear wheel with wide hub.
Brakes:	Front brake 180 mm. Rear brake 180 mm with pressed brake drums.
Speedometer:	Till 7246: 'VDO' with built-in lighting. From 7247 : 'Smiths'. Ratio 1:2 (special fit, see note)
Lights:	Till 7064: Head light 'Riemann' make, with flat, corrugated front glass and double reflector. From 7065: Head light »Lucas« made, single point fixing or side fitted. Polished tail light with a loose partition and a clear red/yellow celluloid plate.
Horn:	'Riemann' or 'Klaxon' (special fit, see note)

Note:

In many ways, the machines 7247 – 7500 are transition versions in different forms. The 'high' front fork was introduced in 1948, but earlier, many aspects pointed in that direction. All headlights were 'Lucas' on the first machines – single point fixed, and later side fitted.

Front forks have been found where the supports for the bottom-fitted headlights were sawn off. The speedometer was 'Smiths' made, fitted onto the support, which was until then, used for the horn, midway on the handlebars, bent up about 10°. The horn can either be a 'Riemann' or, later, a 'Klaxon', fitted at the left-hand side of the frame below the seat, where the clamping plate for guiding the hand operated gear lever, if any, normally is fitted.

1948

Year:	1948 7501 - 8000, in total 500 units.
Versions:	'Special'. All machines from 1947 – 53 were called 'Special'.
Colour:	Black with double gold pin striping
Prices:	Factory price DKK 3,675. Dealer price DKK 4,583.40 including pillion seat.
Number:	Frame- and engine numbers are identical. Make- and frame number (see chapter *Numbers etc.') on a plate (brass, occasionally aluminium or zinc) on the frame at the left hand side below the seat.
Engine:	22 HP, compression 1:5.7. Splash-lubricated crankshaft, drip-lubricated camshaft.
Carburettor:	Nimbus '38 with fuel hose. German rotating fuel valve.
Dynamo:	Dynamo '39, two brush dynamo 6/8 V, 70 W. Voltage regulator '48 under seat.
Gearbox:	Transition model, which means a new model, with the 3rd gear having straight cut teeth. Hand operated clutch.
Transmission:	Cushioned drive shaft. Ratio 12:59.
Fuel tank:	Outlet at the mid left side of the base. Upward reinforcement ribs in the base. Dedicated transfer for 'Special' version.
Frame:	Double head tube. Hole in the baffle plate for cushioned driveshaft. Two holes in the left hand fish plate of the frame for attaching the final drive housing. Centre stand plates with reinforcement. No slot for the gear lever.
Front fork:	Front fork '48 with rubber sleeves. Friction damper with Bakelite handgrip.
Handlebars:	Handlebars for high front fork, 10° bent upward with no hole for the speed-ometer. Flat handlebars in combination with a high front fork are found as well. Round headed screws for combination switch. 28 mm hex centre bolt.
Front mudguard:	Open, without valances. Roundish top side.
Rear mudguard:	Somewhat rounded top side. Straight at the rear end from the tail light down.
Centre stand:	Reinforcement around the hinges. Welded-on feet

Seat:	Supported behind cross plate. Compression coil springs with angled plates.
Pillion seat:	Hinges outside the frame. Compression coil springs with angled plates. Handgrip made from 14 mm chrome plated round steel or brass.
Wheels:	Front wheel with straight spokes. Rear wheel with wide hub.
Brakes:	Front brake 180 mm. Rear brake 180 mm with pressed brake drums with gills.
Speedometer:	'Smiths'. Ratio 1:2
Lights:	Head light 'Lucas', side fitted. Polished tail light with a loose partition and a clear red/yellow celluloid plate.
Horn:	'Klaxon'.

1949

Year:	1949, 8001 - 8825, in total 825 units.
Versions:	'Special'. All machines from 1947 – 53 were called 'Special'.
Colour:	Black with double gold pin striping.
Prices:	Factory price DKK 3,675. Dealer price DKK 4,583.40 including pillion seat.

Number:	Frame- and engine numbers are identical. Make- and frame number (see chapter 'Numbers etc.') on a plate (brass, occasionally aluminum or zinc) on the frame at the left hand side below the seat.
Engine:	22 HP, compression 1:5.7. Till 8500: Flat head gasket. From 8501: Ring shaped head gasket. Splash-lubricated crankshaft, drip-lubricated camshaft.
Carburettor:	Till 8500: Nimbus '38 with fuel hose. From 8501: Nimbus '50. Till about 8500: German rotating fuel valve. From about 8500: Dual English slide fuel valve »Ewarts«.
Dynamo:	Dynamo '39, two brush dynamo 6/8 V, 70 W. Voltage regulator '48 under seat.
Gearbox:	Transition model, which means a new model, with the 3rd gear having straight cut teeth. Hand operated clutch.
Transmission:	Cushioned drive shaft. Ratio 12:59.
Fuel tank:	Outlet at the mid left side of the base. Upward reinforcement ribs in the base. Dedicated transfer for 'Special' version.
Frame:	Double head tube. Hole in the baffle plate for cushioned driveshaft. Two holes in the left hand fish plate of the frame for attaching the final drive housing. Centre stand plates with reinforcement.
Front fork:	Front fork '48 with rubber sleeves. Friction damper with Bakelite handgrip.
Handlebars:	Handlebars for high front fork, 10° bent upward with no hole for the speedometer. Flat handlebars in combination with a high front fork are found as well. Round headed screws for combination switch. 28 mm hex centre bolt.
Rear mudguard:	Somewhat rounded top side. Straight at the rear end from the tail light down.
Front mudguard:	Till 8800: Open, without valances. Roundish top side. From 8801: With low valances (in one piece) or with rolled-on valances. Roundish top side.
Centre stand:	Reinforcement around the hinges. Welded-on feet
Seat:	Supported behind cross plate. Compression coil springs with angled plates.
Pillion seat:	Hinges outside the frame. Compression coil springs with angled plates. Handgrip made from 14 mm chrome plated round steel or brass.

Wheels:	Front wheel till 8500 with straight spokes. From 8501 with bent spokes. Rear wheel with wide hub.
Brakes:	Front brake 180 mm. Rear brake 180 mm with pressed brake drums with gills.
Speedometer:	'Smiths'. Ratio 1:2
Lights:	Till appr. 8000: Head light 'Lucas', side fitted. From appr. 8000: Head light 'Hella' without speedometer. Polished tail light with a loose partition and a clear red/yellow celluloid plate.
Horn:	'Klaxon'.

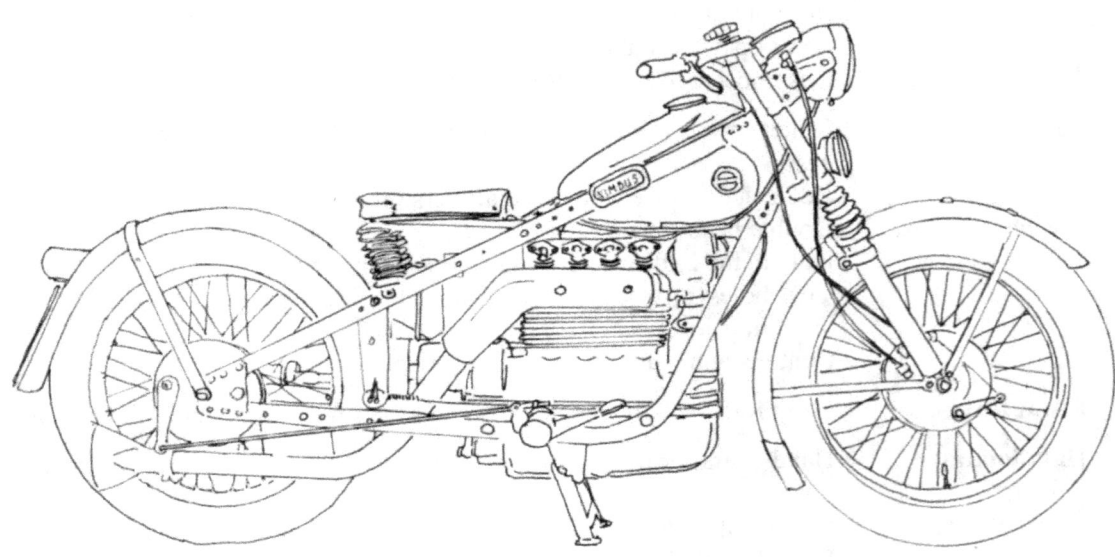

1950 – 51

Year:	1950 – 51, 8826 - 10399, in total 1574 units.
Versions:	'Special'. All machines from 1947 – 53 were called 'Special'.
Colour:	Black with double gold pin striping.
Prices:	Factory price Nov. 1951: DKK 3,700 Dealer price DKK 4,931
Number:	Frame- and engine numbers are identical. Make- and frame number on a plate (brass, occasionally aluminum or zinc) on the frame at the left hand side below the seat.
Engine:	22 HP, compression 1:5.7. Ring shaped head gasket. Splash-lubricated crankshaft, drip-lubricated camshaft.
Carburettor:	Till 9200: Nimbus '50. From 9201 – 9630: Nimbus 51-1. From 9631: Nimbus '51-2. Till appr. 9000: dual English slide fuel valve. From appr. 9000: German rotating fuel valve.
Dynamo:	Till appr. 10000: dynamo '39. From appr. 10000: dynamo '51. Two brush dynamo 6/8 V, 70 W. Voltage regulator '48 under seat.
Gearbox:	Till 9000 transition model, which means a new model, with the 3rd gear having straight cut teeth. From 9001: late model with the 3rd gear having angled cut teeth. Till 8630: foot operated gear shift without neutral gear indicator and with a round boss for clutch cable fixing From 8631: foot operated gear shift with neutral gear indicator and angular boss (with a hole for clutch cable if needed for the early gearbox)
Transmission:	Cushioned drive shaft. Ratio 12:59.
Fuel tank:	Outlet at the mid left side of the base. Upward reinforcement ribs in the base. Dedicated transfer for 'Special' version.
Frame:	Double head tube. Hole in the baffle plate for cushioned driveshaft. Two holes in the left hand fish plate of the frame for attaching the final drive housing. Centre stand plates with reinforcement.
Front fork:	Till 9000: Front fork '48. From 9001: front fork '50. With rubber sleeves. Friction damper with Bakelite handgrip.
Handlebars:	Handlebars for high front fork, 10° bent upward with no hole for the speedometer.

	Round headed screws for combination switch.
	28 mm hex centre bolt.
Front mudguard:	With low valances (in one piece) or with rolled-on valances. Roundish top side.
Rear mudguard:	Till 9000: Somewhat rounded top side. Straight at the rear end from the tail light down.
	From 9001: Flat topside and bent end with a bracket for the number plate.
Centre stand:	Reinforcement around the hinges.
	Welded-on feet
Seat:	Till 9000: Supported behind cross plate. Compression coil springs with angled plates.
	From 9001: Supported by the mid section of the cross plate in rubber bushings and at the rear in heavy gauge rubber bands.
Pillion seat:	Hinges outside the frame.
	Till 9000: Compression coil springs with angled plates.
	From 9001: Front pivots in rubber bushings, at the rear in heavy gauge rubber bands. Handgrip made from 14 mm chrome plated round steel or brass for the 'Luxus' and PVC covered for the 'Standard' version.
Wheels:	Front wheel with bent spokes.
	Rear wheel with wide hub.
Brakes:	Front brake 180 mm.
	Rear brake 180 mm with pressed brake drums with one wide gill.
Speedometer:	Till around 9500:'Smiths'. From around 9500: 'VDO' 80 mm, fitted in the head light. Ratio 1:2.
	Military machines were always from then off fitted with "Smiths" speedometers.
Lights:	Till appr. 9500: Head light 'Hella' without speedometer, side fitted.
	From appr. 9500: Head light 'Hella' with 80 mm 'VDO' speedometer.
	Polished tail light with a loose partition and a clear red/yellow celluloid plate.
Horn:	Till around. 9000: 'Klaxon'.
	From around 9000: 'Hella'.

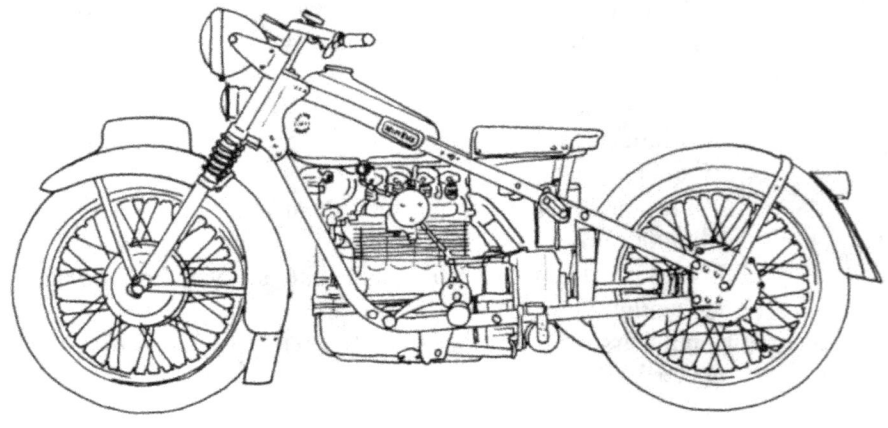

1952 – 53

Year:	1952 – 53, 10400 – 12178, 12180 – 12190 and 12212 - 12223, in total 1802 units.
Versions:	'Special'. All machines from 1947 – 53 were called 'Special'.
Colours:	Black or red with gold pin striping
Prices:	With licence (note) Factory price: DKK 4,620. Dealer price DKK 6,228 including pillion seat
Number:	Frame- and engine numbers are identical. Make- and frame number on a plate (brass, occasionally aluminum or zinc) on the frame at the left hand side below the seat.
Engine:	Till 11100: 22 HP, compression 1:5.7. Ring shaped head gasket. From 11101: 18 HK, compression 1:5. Splash-lubricated crankshaft, drip-lubricated camshaft.
Carburettor:	Till 11300: Nimbus '51-2. From 11301: Nimbus '53. German rotating fuel valve.
Dynamo:	Dynamo '51. Two brush dynamo 6/8 V, 70 W. Voltage regulator '48 under seat.
Gearbox:	Late model with the 3rd gear having angled cut teeth. Foot operated gear shift with neutral gear indicator and angular boss (with a hole for clutch cable if needed for the early gearbox)
Fuel tank:	Outlet at the mid left side of the base. Upward reinforcement ribs in the base. Dedicated transfer for 'Special' version.
Frame:	Double head tube. Till 10440: Thin cross bracket. From 10441: Heavy gauge cross bracket. Till 11970: Flat clamping plate. From 11971: Clamping plate with bent edges. Hole in the baffle plate for cushioned driveshaft. Two holes in the left hand fish plate of the frame for attaching the final drive housing. Centre stand plates with reinforcement.
Transmission:	Cushioned drive shaft. Ratio 12:59.
Front fork:	Front fork '50. With rubber sleeves. Friction damper with Bakelite hand-grip.
Handlebars:	Handlebars for 'high' front fork, 10° bent upward with no hole for the speedometer. Round headed screws for combination switch. 28 mm hex centre bolt.
Front mudguard:	With low valances (in one piece) or with rolled-on valances. Roundish top side.

Rear mudguard:	Flat topside and bent end with a bracket for the number plate.
Centre stand:	Reinforcement around the hinges. Welded-on feet
Seat:	Supported by the mid-section of the cross plate in rubber bushings and at the rear in heavy gauge rubber bands.
Pillion seat:	Front pivots in rubber bushings, at the rear in heavy gauge rubber bands. Handgrip made from 14 mm chrome plated round steel or brass for the 'Luxus' and PVC covered for the 'Standard' version.
Wheels:	Front wheel with bent spokes. Rear wheel with wide hub.
Brakes:	Front brake 180 mm. Rear brake 180 mm with pressed brake drums with one wide gill.
Speedometer:	"VDO" 80 mm, fitted in the head light. Ratio 1:2. Military machines were fitted with "Smiths" speedometers.
Lights:	Head light "Hella" with 80 mm "VDO" speedometer. Polished tail light with a loose partition and a clear red/yellow celluloid plate.
Horn:	"Hella".

Note:

In the early nineteen-fifties is was impossible to buy a new vehicle, unless one had a licence from the Government. The reason was Denmark's exchange rate problem; most raw materials and equipment had to be purchased from abroad. A licence was issued in those cases where there was a real need to own a vehicle, for example midwives, doctors, architects and for some employees working for utility companies. Needless to mention was that at that time second hand Nimbuses were very expensive.

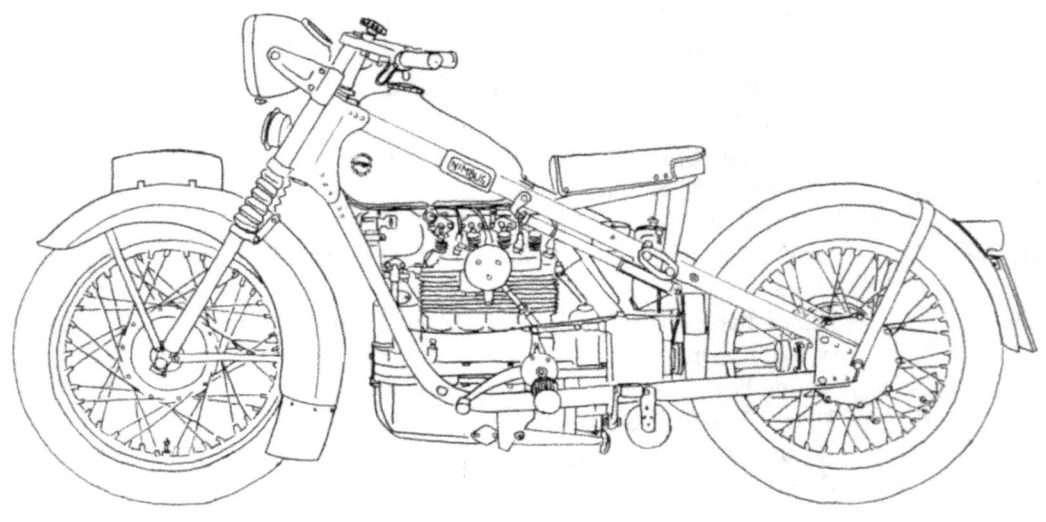

1954 – 55

Year:	1954 – 55, 12179, 12192 - 12211 and 12224 - 13572, in total 1370 units.
Versions:	'Standard' and 'Luxus'
Colours:	'Standard': Black with double gold pin striping.
	'Luxus': Black, red, withered green or deep sea green with double gold pin striping
Prices:	Factory price 'Standard': DKK 4,395, 'Luxus' DKK 4,620..
	Dealer price 'Standard' DKK 5,913, 'Luxus' DKK 6,228 including pillion seat
Number:	Frame- and engine numbers are identical.
	Make- and frame number on a plate (brass, occasionally aluminum or zinc) on the frame at the left hand side below the seat.
Engine:	22 HP, compression 1:5.4. Ring shaped head gaskets.
	Splash-lubricated crankshaft, drip-lubricated camshaft.
Carburettor:	Nimbus '53.
	German rotating fuel valve.
Dynamo:	Dynamo '51. Two brush dynamo 6/8 V, 70 W.
	Till 12600: Voltage regulator '48 under seat.
	From 12601: 'Bosch' voltage regulator under seat.
Gearbox:	Late model with the 3rd gear having angled cut teeth.
	Foot operated gear shift with neutral gear indicator and angular boss (with a hole for clutch cable if needed for the early gearbox)
Transmission:	Cushioned drive shaft. Ratio 12:59 or 14:57
Fuel tank:	Outlet at the mid left side of the base.
	'Luxus': Enameled tank badges.
	'Standard': transfer/decal 50 mm.
	Till 12200 upward reinforcement ribs in the base.
	From 12201 upward reinforcement ribs in the base and sides.
Front fork:	Front fork '50. With rubber sleeves. Friction damper with Bakelite hand-grip.
Frame:	Till 13040: Double head tube with riveted-on headstock shield.
	From 13041: Single or double head tube with welded-on headstock shield and triangular shaped reinforcement between bottom brackets.
	Hole in the baffle plate for cushioned driveshaft.
	Two holes in the left hand fish plate of the frame for attaching the final drive housing.
	Centre stand plates with reinforcement.
Handlebars:	Handlebars for high front fork, 10° bent upward with no holes for the instruments.
	Round headed screws for combination switch.
	28 mm hex centre bolt.

Front mudguard:	With low valances (in one piece) or with rolled-on valances. Roundish top side.
Rear mudguard:	Till 12500: Flat topside and rounded end with a bracket for the number plate.
	From 12501: Rounded topside and roundish end with a bracket for the number plate.
Centre stand:	Reinforcement around the hinges.
	Welded-on feet
Seat:	Supported by the mid-section of the cross plate in rubber bushings and at the rear in heavy gauge rubber bands.
Pillion seat:	Front pivots in rubber bushings, at the rear in heavy gauge rubber bands. Handgrip made from 14 mm chrome plated round steel or brass for the 'Luxus' and PVC covered for the 'Standard' version.
Wheels:	Front wheel with bent spokes.
	Rear wheel with wide hub.
Brakes:	Front brake 180 mm. Rear brake 180 mm with pressed brake drums with one wide flange.
Speedometer:	'VDO' 80 mm, fitted in the head light. Ratio 1:2.
	Military machines were fitted with 'Smiths' speedometers.
Lights:	Head light 'Hella' with 80 mm 'VDO' speedometer.
	On military machines: 'Lucas' without speedometer.
	Till 12600: Tail light with bottom plate of thin black rubber.
	From 12601: Tail light with bottom ring of thick grey rubber.
	Till 13200: Polished tail light with a loose partition and a clear red/yellow celluloid plate.
	13201 – 13600: Red glass reflector and chrome plated ring.
	From 13601: Red plastic reflector, J.R.U 129.

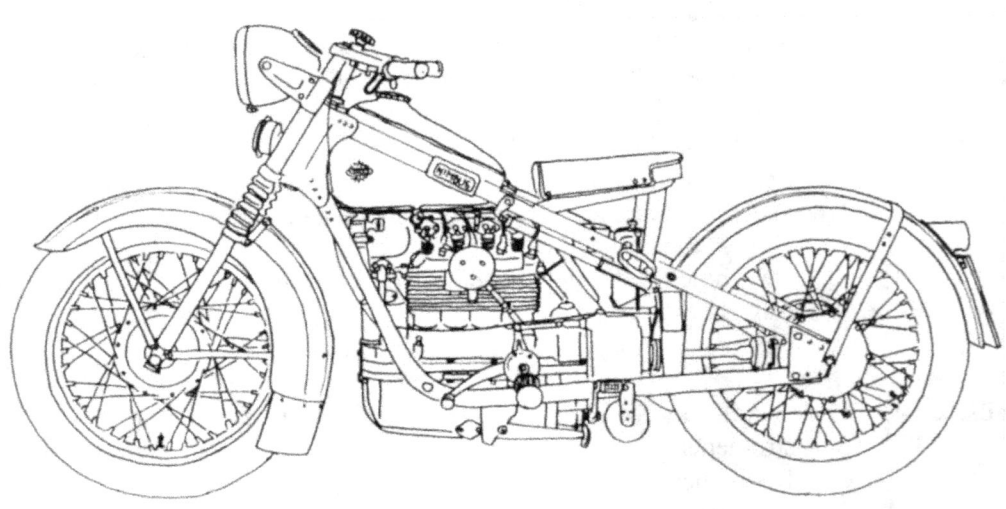

1956

Year:	1956, 13573 – 13769, 13775 - 13777 and 13801 - 13854, in total 254 units.
Versions:	'Standard' and 'Luxus'.
Colours:	Black, red, withered green or deep sea green with double gold pin striping
Prices:	Factory price 'Standard': DKK 4,650, 'Luxus' DKK 4,800. Dealer price 'Standard' DKK 6,270, 'Luxus' DKK 6,480 + 'Dollar grant charge' (See note) DKK 1,050, in total DKK 7,320 for the 'Standard' and DKK 7,530 for the 'Luxus'.
Number:	Frame number at the left hand side of the headstock shield. Frame- and engine numbers (= production numbers) are different. (The frame numbers were not issued in a logical sequence) Make, production- and engine number on an aluminum plate on the frame at the left hand side below the seat.
Engine:	22 HP, compression 1:5.4. Ring shaped head gaskets. Till 15372: Open valves. From 15573: Fully enclosed valve mechanism, except for machines for military- and postal services. Splash-lubricated crankshaft, drip-lubricated camshaft.
Carburettor:	Nimbus '53. German manufacture rotating fuel valve.
Dynamo:	Dynamo '51. Two brush dynamo 6/8 V, 70 W .'Bosch' voltage regulator under seat.
Gearbox:	Late model with the 3rd gear having angled cut teeth. Foot operated gear shift with neutral gear indicator and angular boss (with a hole for clutch cable if needed for the early gearbox)
Fuel tank:	Outlet at the mid left side of the base. 'Luxus': Enameled tank badges. 'Standard': transfers/decals 50 mm. Reinforcement ribs in the base and sides.
Frame:	Till 13572: Single or double head tube with welded-on headstock shield and triangular shaped reinforcement between bottom brackets. From 13573: Double head tube with welded-on headstock shield, with make, frame number, handlebars lock, and triangular shaped reinforcement between bottom brackets. Hole in the baffle plate for cushioned driveshaft. Two holes in the left hand fish plate of the frame for attaching the final drive housing. Centre stand plates with reinforcement.

Front fork:	Front fork '50. With rubber sleeves. Friction damper with Bakelite handgrip.
Handlebars:	Handlebars for high front fork, 10° bent upward with no holes for the instruments. Round headed screws for combination switch. 28 mm hex centre bolt.
Front mudguard:	With low valances (in one piece) or with rolled-on valances. Roundish top side.
Rear mudguard:	Rounded topside and roundish end with a bracket for the number plate. Stays attached with rivets in a recess across the mudguard have been found.
Centre stand:	Reinforcement around the hinges. Welded-on feet
Seat:	Supported by the mid-section of the cross plate in rubber bushings and at the rear in heavy gauge rubber bands.
Pillion seat:	Front pivots in rubber bushings, at the rear in heavy gauge rubber bands. Handgrip made from 14 mm chrome plated round steel or brass for the 'Luxus' and PVC-covered for the 'Standard' version.
Wheels:	Front wheel with bent spokes. Rear wheel with wide hub.
Brakes:	Front brake 180 mm. Till 13850: Front brake anchor plate with short torsion section against fork leg. From 13851: Front brake anchor plate with long torsion section against fork leg. Rear brake 180 mm with cast brake drum with one wide flange.
Speedometer:	'VDO' 80 mm, fitted in the head light. Ratio 1:2. Military machines were fitted with 'Smiths' speedometers.
Lights:	Head light 'Hella' with 80 mm 'VDO speedometer. Tail light: : From 12601: Tail light with bottom ring of thick grey rubber. 13201 – 13600: Red glass reflector and chrome plated ring. From 13601: Red plastic reflector, J.R.U 129.
Horn:	'Hella'.

Note:

A Dollar grant charge is an extra charge to be paid to the Government in case one was able and allowed to buy a Nimbus with US Dollars. In those cases where a licence as described in the note on the 1952-'53 was not granted.

1957 – 59

Year:	1957 - 59 , 13770 – 13774, 13778 - 13800 and 13855 - 14015 - 13572, in total 189 units.
Versions:	'Standard' and 'Luxus'
Colours:	Black, red, withered green or deep sea green with double gold pin striping
Prices:	Factory price 'Standard': DKK 4,650, 'Luxus' DKK 4,800. Dealer price after July 6, 1957: 'Standard' DKK 8,362.50, 'Luxus' DKK 8,640 including pillion seat

Number	Frame number at the left hand side of the headstock shield.
	Frame- and engine numbers (production numbers) are different.
	Make, production- and engine number on an aluminum plate on the frame at the left hand side on the frame below the seat.
Engine:	22 HP, compression 1:5.4. Ring shaped head gaskets.
	Fully enclosed valve mechanism, except for machines for military- and postal services.
	Splash-lubricated crankshaft, drip-lubricated camshaft.
Carburettor:	Nimbus '53.
	German manufacture rotating fuel valve.
Dynamo:	Dynamo '51. Two brush dynamo 6/8 V, 70 W. "Bosch" under seat.
Gearbox:	Late model with all three gears having angled teeth.
	Foot operated gear shift with neutral gear indicator and angular boss (with a hole for clutch cable if needed for the early gearbox)
Transmission:	Cushioned drive shaft.
	Ratio 12:59 (low geared) or 14:57 (high geared)
Fuel tank:	Outlet at the mid left side of the base.
	'Luxus': Enameled tank badges.
	'Standard': transfer/decal 50 mm.
	Reinforcement ribs in the base and sides.
Frame:	Double head tube with welded-on headstock shield with make and frame number, handlebars steering lock, and triangular shaped reinforcement between bottom brackets.
	Hole in the baffle plate for cushioned driveshaft.
	Two holes in the left hand fish plate of the frame for attaching the final drive housing.
	Centre stand plates with reinforcement.
Front fork:	Front fork '50. With rubber sleeves. Friction damper with Bakelite handgrip.
Handlebars:	Handlebars for high front fork, 10° bent upward with no holes for the instruments.
	Round headed screws for combination switch.
	28 mm hex centre bolt.
Front mudguard:	With low valances (in one piece) or with rolled-on valances. Roundish top side.
Rear mudguard	Rounded topside and roundish end with a bracket for the number plate.
	Stays attached with visible rivets in a recess across the mudguard.
Centre stand:	Reinforcement around the hinges.
	Welded-on fee
Seat:	Supported by the mid-section of the cross plate in rubber bushings and at the rear in heavy gauge rubber bands.
Pillion seat:	Front pivots in rubber bushings, at the rear in heavy gauge rubber bands.

	Handgrip made from 14 mm chrome plated round steel or brass for the 'Luxus' and PVC covered for the 'Standard' version.
Wheels:	Front wheel with bent spokes.
	Rear wheel with wide hub.
Brakes:	Front brake 180 mm.
	Front brake anchor plate with long torsion section against fork leg.
	Rear brake 180 mm with cast brake drum with one wide flange.
Speedometer:	'VDO' 80 mm, fitted in the head light. Ratio 1:2.
Lights:	Head light 'Hella' with 80 mm 'VDO' speedometer.
	Tail light with bottom ring of thick grey rubber.
	Red plastic reflector, J.R.U 129.
Horn:	'Hella'.

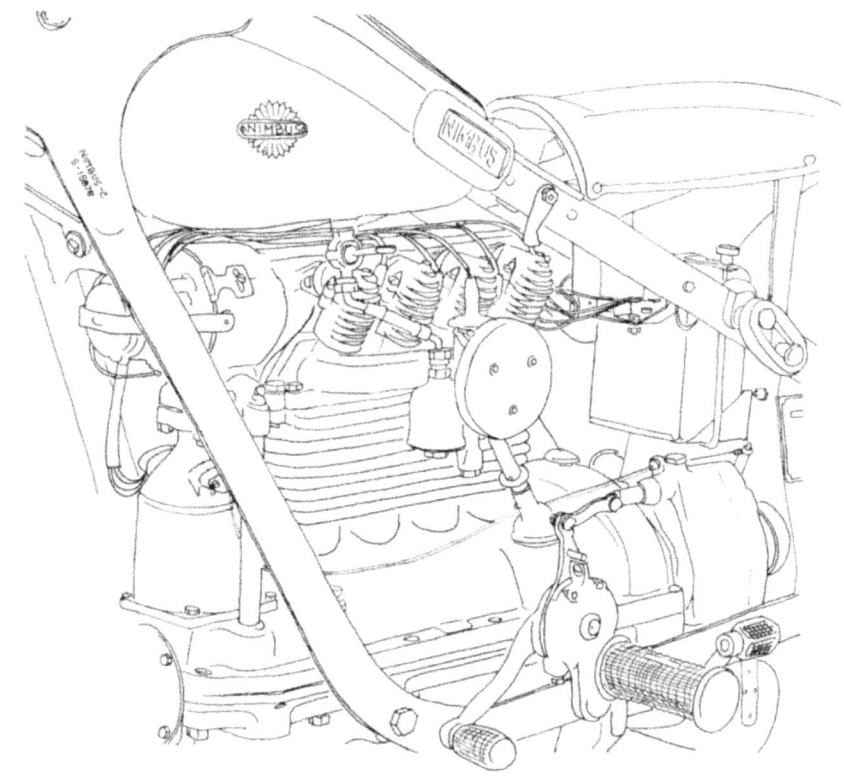

FRAME

The Nimbus-C frame is patented and has basically been used without major changes from 1934 till 1960. The majority of the frame is made from 8x40 mm flat steel. It is constructed like a cradle **(1)** with a right and a left hand frame rail **(2)**. The head tube **(3)** is fitted at the front, while the frame plate **(4)** is placed halfway along the upper frame. Between the rearward upper and lower frame rails, a baffle plate **(5)** is fitted which bears the battery carrier **(6)**.

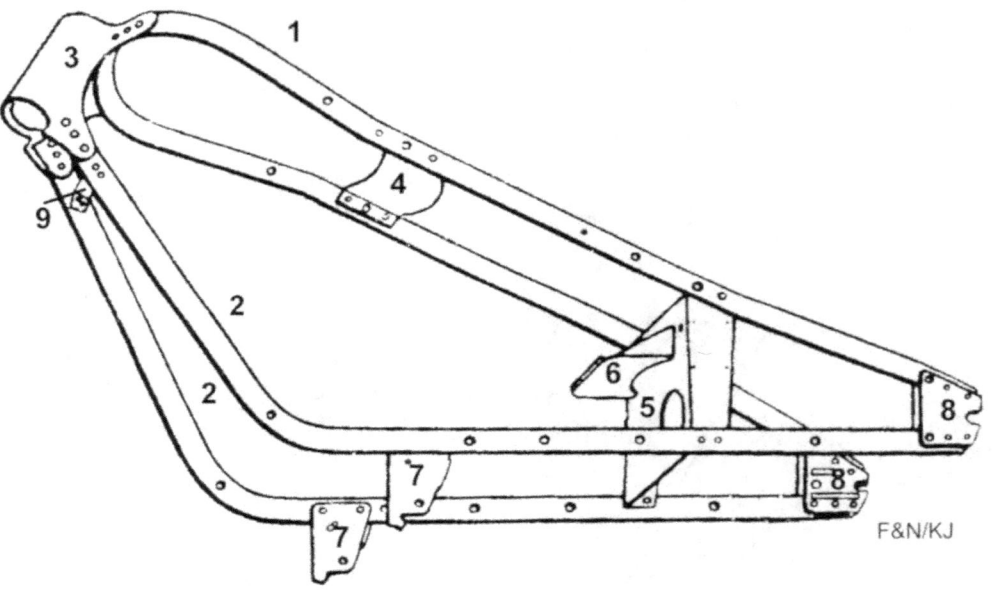

The centre stand pivot plates **(7)** are fitted to the lower frame rails, as are the rear left and right hand axle fish plates **(8)** which have guiding grooves for the final drive housing, on their inner faces, and make the joint to the upper rails.
Finally, a cross bracket **(9)** is found right under the head tube between the frame rails.

Although the frame was basically the same during the years, a lot of changes took place.

Frame '34, 1301 – 1550
The head tube of the first frames in 1934 consisted of a single tube, which was attached to the cradle frame by means of a pair of front flanges and to the frame rails with a band shaped tube that surrounded the lower part of the head tube. There was one hole in the baffle plate to accommodate the rigid driveshaft and one hole in the centre of the left hand fish plate for attaching the final drive housing. The battery carrier was riveted to the baffle plate. There were no holes for attaching the voltage regulator, but one extra hole was drilled at the right hand side for the exhaust heat shield. The two centre stand pivot plates were short, and the cross bracket (9) was frail. There was a slot in the frame plate (4) for the hand-change gear lever and an angle support for the clutch cable attached to the left frame rail.

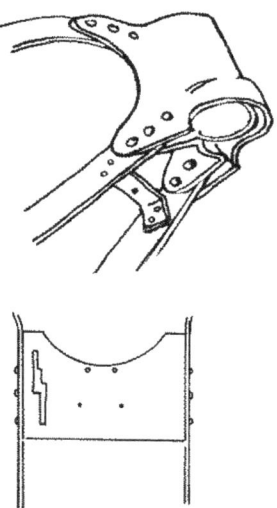

Frame '35, 1551 – 2560
To strengthen the connection between the head tube and the frame, a flanged 'shield' was fitted that surrounded the head tube over its full length. In addition, the frame plate was provided with holes for fitting the clamping plate for the fuel tank, instead of the tiny '34 clamp.

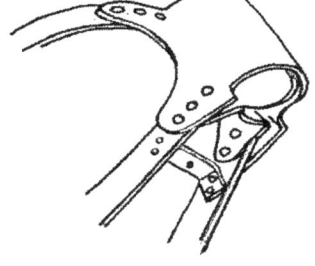

Frame '36, 2561 – 2900
The rigid driveshaft was replaced by the cushioned driveshaft, which had a bigger diameter than the rigid version. Therefore, the hole in the baffle plate was made accordingly bigger. The left rear fish plate was changed; the hole in the middle became obsolete, and two holes for attaching the final drive housing were drilled correspondingly in the cradle frame, fish plate and frame rail at the places where there were two rivets before.

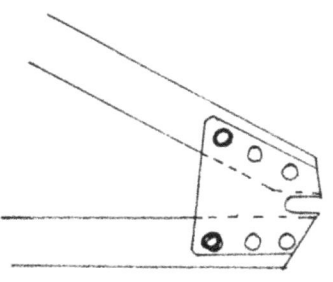

Frame '37, 2901 – 7500
The centre stand was changed (see chapter *Centre stand*) and consequently the centre stand pivot plates were also changed to make this new centre stand fit.

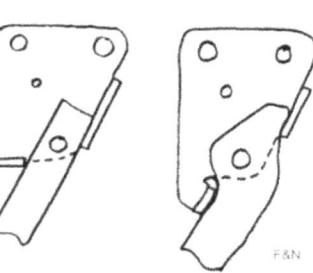

A hole was drilled in the left hand frame rail near the baffle plate for the pivot for the foot-change gear pedal. Nevertheless, the two holes for pivoting the hand-change gear lever remained in the frame plate. This made it possible to fit either a hand-change gear lever or a foot-change gear pedal for all machines from 2901 – 7500.

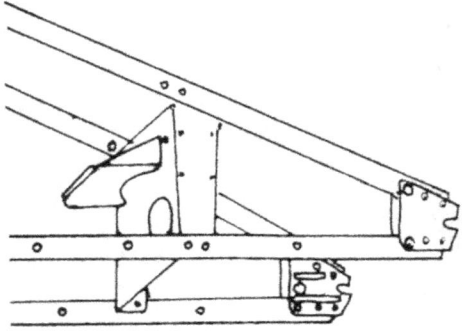

Frame '48, 7501 – 9000
As the hand-operated gear shift became obsolete, the slot in the frame plate for the hand-change gear lever was left out. The small angle for the clutch cable at the left hand frame rail was omitted as well. The battery carrier was moved somewhat higher up and to the right on the baffle plate, to make room for the operating lever of the late gear box. The battery carrier was now welded to the baffle plate, instead of riveted.

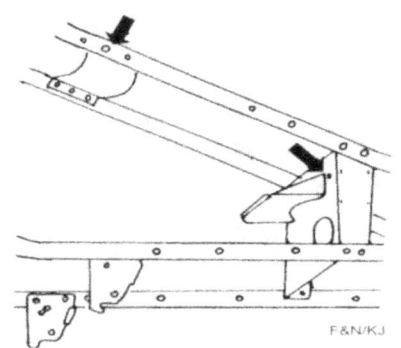

Frame '50, 9001 –10440
The seat and the pillion seat were now sprung in heavy gauge rubber bands instead of in compression coil springs.
This means that the holes for the hinges of the heavy gauge rubber bands were situated just in front of the baffle plate, and that the pivot of the seat was placed internally below the baffle plate, where in older models the middle rivet was situated.

The front air pump angle bracket at the left hand side of the upper frame rail was moved forward, and instead of a rear angle bracket, a small support pin was fitted in the baffle plate.

Until now, the front clamping plate for the fuel tank was attached to threaded studs between frame and head tube. That construction was abandoned and a bracket, secured by M6 screws at the underside of the head tube, was fitted instead.

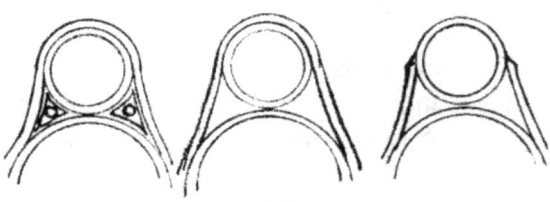

Frame '52, 10441 – 11970

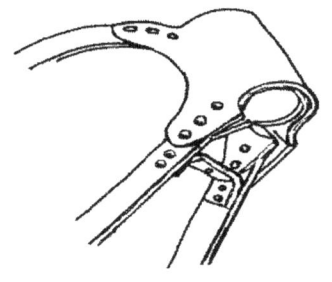

The cross bracket under the head tube was improved to become of more robust construction and was fitted differently in order to minimize torsion, especially when riding with a sidecar attached. From October 1953, this new cross bracket was attached to the frame with larger (8 mm) rivets.

Frame '53, 11971 – 13040

The frame plate between seat and fuel tank was changed. The fuel tank was kept in place at the rear end either with a clamp just like at the front end, or with a clamping plate of a new construction, that had edges folded to fit over the frame rails.

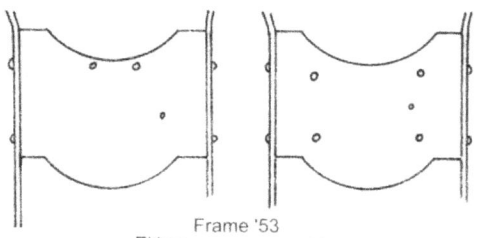

Frame '53
Either - - - - - - - - - - or

Frame '55, 13041 – 13572

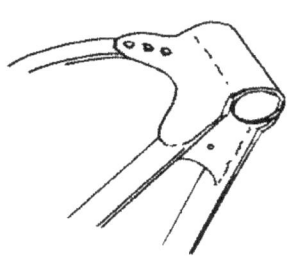

A distinctive change was made in the way the head tube was fitted. From now on, the head tube shield was welded onto the frame rails. Between the frame rails, a triangular shaped plate was welded instead of the cross bracket used before. The way this was carried out was done either by welding a shield all the way round the head tube, or with welded-on head tube flanges.

Frame '56, 13572 (frame number-15001) – 14015

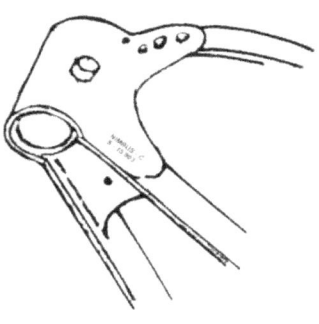

From April 1st, 1956, Denmark was facing a series of new regulations for motorized vehicles. New vehicles from that date onward must have the frame number engraved in the frame. For the Nimbus-C the position was at the left hand flange of the head tube shield. By the way, the frame- and production number (= engine number) were no longer the same. The welded-on head tube shield with cross plate made it possible to guide the wiring to the handlebars between the head tube and the tank. Therefore the front tank clamp was split into two halves. Both halves were kept in place by means of an angled bolt, hooked together in a hole in the edge of the head tube shield. At the same time a cylinder-type security lock was placed at the left hand side of the head tube shield. There are two extra holes in the frame plate, to give access to the rear bolts of the camshaft cover, which at the same time became included on the cylinder head.

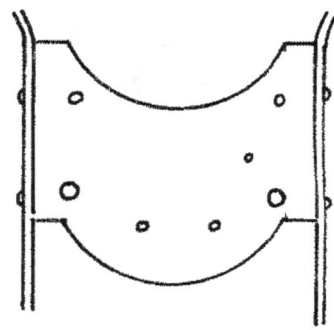
In determining the year of production of the frame, one has to be aware of the fact that frame repairs can have taken place in the course of time, where parts were used without bothering about originality. The number of frames that left the factory is much greater than the 12715 frames that are recorded in the stock books. Many more frames were used, especially, in the first instance by the army.

CENTRE STAND

Centre stands for Nimbus-C come in three versions:

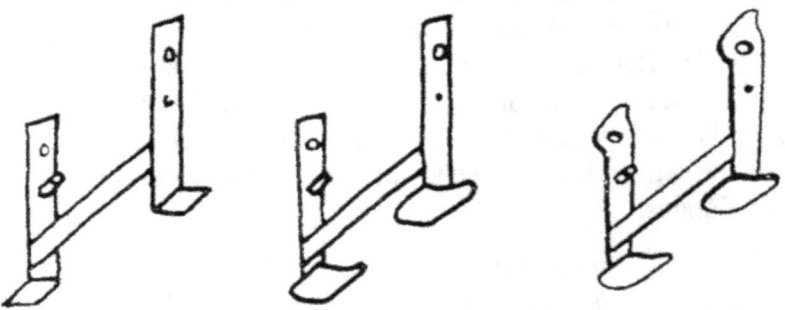

Centre stand 1301 – 1550
Stand made from bent and welded flat steel (5 x 25 mm) with one-piece 'feet' sections bent outwards.

Centre stand 1551 – 2900
As 1301 – 1550, but with the feet sections welded-on.

Centre stand 2901 – 14015
A more robust and stable version of the side rails, which necessitated changes to the frame's centre stand pivot plate.

FOOT RESTS

All foot rests for Nimbus-C are the same. The construction is as such, that they can be adjusted to suit the length of the driver's legs (within certain limits). Often, but not always, the footrest rubbers have the »NIMBUS« mark embedded in them. On footrests where the part entering the footrest rubber is rectangular, these are (non-Nimbus) copies. Non-Nimbus fold-up footrests for the passenger were on sale too.
There are different versions of footrest rubbers:

Footrest rubbers 1301 – 9000
Without special marks, except for a fine cross- or line pattern and with convex ends.

Footrest rubbers 9001 – 14015
Marked in the same way the hand grip rubbers are marked, with the »NIMBUS« name moulded-in.

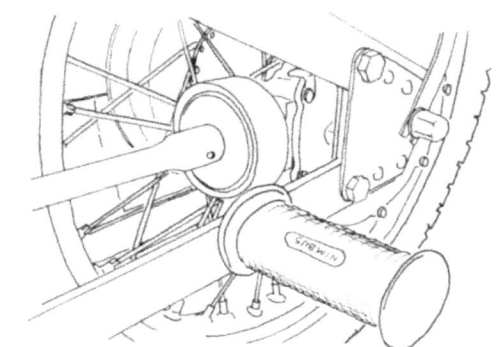

KNEE RUBBERS

The knee rubbers for Nimbus-C remained unchanged during the entire production period. They are made from rubber with the »NIMBUS« name moulded-in and with various casting marks or – numbers inside. A porous insert is fitted inside the knee rubbers, and then onto a sheet iron holder. The holder has four holes to allow the knee rubbers to be fitted to suit the rider's leg length.

TOOL BOX

All tool boxes for Nimbus-C are the same.
Tool boxes for military machines are fitted with a special bracket onto the rear mudguards brace.
(See Military accessories.)

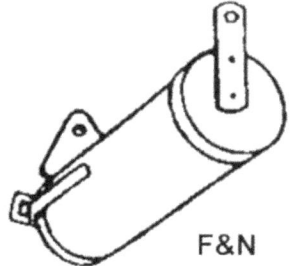

F&N

DRIVER- AND PILLION SEAT

The driver and pillion seat for Nimbus-C come in two main versions:

1301 – 9000: compression coil sprung.

9001 – 14015: heavy gauge rubber band sprung.

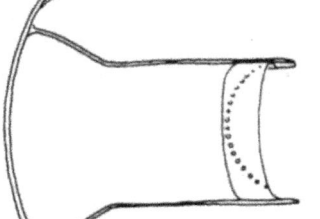

Driver Seat 1301 – 1900
Seat frame made from flat steel (5x25mm) pivoted right behind the frame's cross plate and sprung by two compression coil springs with eyes, either with a 5 mm or a 5.5 mm wire thickness. The front holes for the 20 saddle internal coil springs are 6 mm, the rear ones 4 mm.

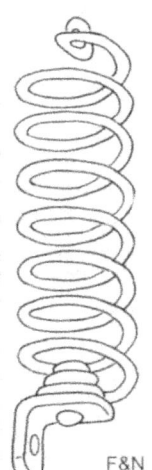

Driver Seat 1901 – 2500
As 1301 – 1900, except that the compression coil springs are attached to a steel angle bracket instead of having an eye direct to the frame.

Driver Seat 2501 – 9000
As 1901 – 2500, except that the 20 rear holes for the saddle internal coil springs are drilled up to 6 mm to allow the springs to move more freely to prevent kinking.

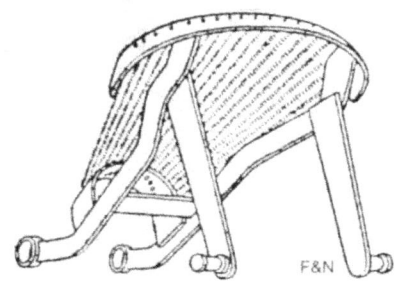

Driver Seat 9001 – 14015
Pivoted mid-under the cross plate of the frame and sprung by two heavy gauge rubber bands attached to two pins welded onto support rods at both sides of the seat. To damp the downward movement, two rubber damping blocks were fitted onto the frame.

Driver Seat prior to 9001?
A few machines prior to 9001 were fitted with a seat sprung with heavy gauge rubber bands, but pivoted behind the frame plate as an *'intermediate'* version. The quantity and production numbers are unknown.

Pillion seat 1301 – 1550
Pillion seat frame made from flat steel (5 x 25 mm), pivoted at the front in the reduced diameter of the enlarged ends of the seat hinge pivot support and sprung with two compression coil springs with a wire thickness of 5.0 or 5.5 mm. The springs with eyes are bolted to the rear mudguard braces. The pillion seat is fitted inside the frame rails. The steel hand grip has the same dimension as the flat steel of the pillion seat frame and is riveted to this frame.

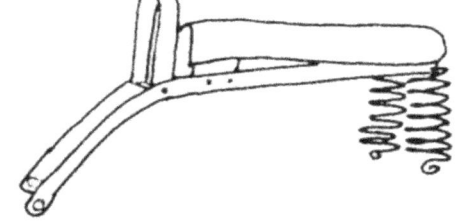

Pillion seat 1551 – 1900
Fitted outside the frame rails with bushes around a pivot. The handgrip is made from chrome plated steel with a diameter of 14 mm and is fitted on a riveted cross bracket. The handgrip is relatively high and narrow. Furthermore, the pillion seat is stabilized with an extra cross bracket, riveted between hand grip and pivot.

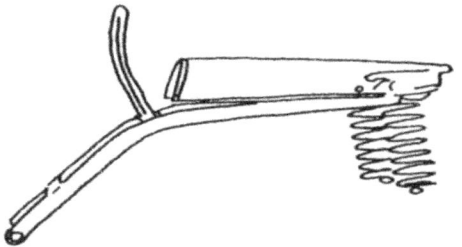

Pillion seat 1901 – 9000
As 1551 – 1900, but welded instead of riveted. The compression coil springs have an angled section to attach them to the mudguard stay.

Pillion seat 9001 – 14015
Pivots in rubber bushings around the rear mudguard's pivot rod, and sprung by two heavy gauge rubber bands each acting around three pins. There is one pin on both welded-on support rails of the pillion seat, and two on the supports bolted onto the rear mudguard brace. One pin on the mudguard support is a retaining pin, while the other one, around which the heavy gauge rubber band is folded, is a stop pin which damps the movement and causes the pillion seat to swing back after compression.

The 'Luxus' and 'Special' versions have a chrome plated hand grip; the one on the 'Standard' (from 1954) is covered with black PVC. Compared to the hand grip for the pillion seat, sprung with compression coil springs, the hand grip is relatively low and wide.

Not all Nimbus-C machines were delivered with a pillion seat as extra VAT had to be paid when owning a motor cycle with a pillion seat. For this reason, in many cases the original registration - and tax document indicate if the machine is provided with a pillion seat or not.

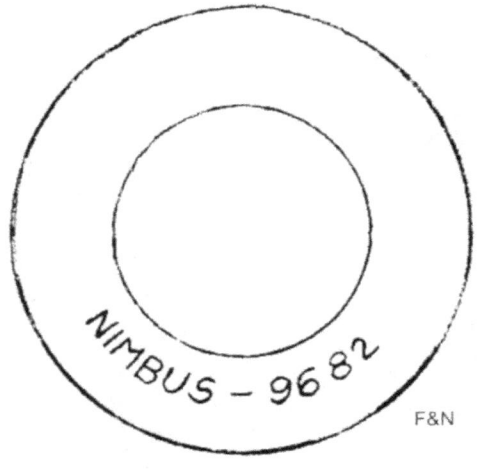

Heavy gauge rubber bands.
The heavy gauge rubber bands for Nimbus-C have the »NIMBUS« name and a 'maximum load' of 80 or 100 kg moulded into them.

Seat cover
The driver and pillion seats were covered in black hide leather. Later models would be provided with synthetic covers.

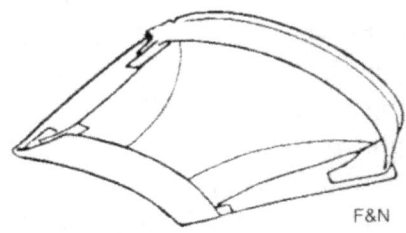

FUEL TANK

Nimbus-C fuel tanks basically have the same shape on all machines.
There are however four versions:

Fuel tank 1301 – 2050
Flat base with the fuel valve outlet at the front left side of the base.

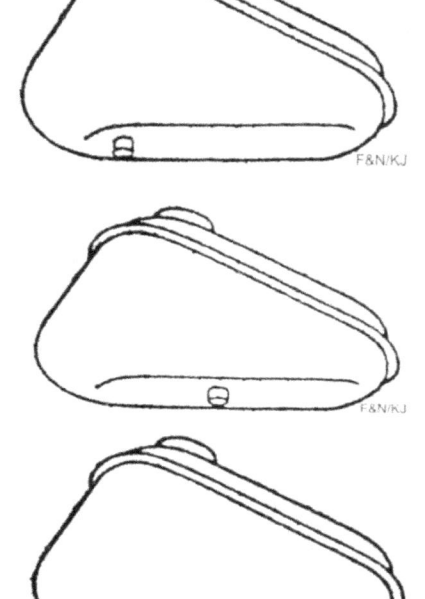

Fuel tank 2051 – 7500
As for 1301 – 2050, but with the fuel valve outlet at the mid left side of the base.

Fuel tank 7501 – 12200
Upward reinforcement ribs in the base to reduce the risk of tears in the base due to material tensions caused by the heat of the engine.

Fuel tank 12201 – 14015
As 7501 – 12200, but for the 'Luxus' version at both sides two small threaded studs for fitting the enamel name badge.
In addition, there is an embossing on both sides under the cradle frame, which, just like the base ribs, serves, to make the material more flexible to reduce the likelihood of failure.

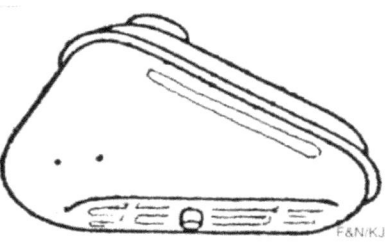

Stainless steel fuel tank
In spite of the reinforcement ribs and embossed sections, many fuel tanks would leak in the course of time. Because repairs are difficult, many owners have chosen to replace the original fuel tank with a new one made from stainless steel.

Rubber ring 2015 - 14015
A rubber ring for fitting between frame and fuel tank was introduced from 2051. This ring, which protects the fuel tank, can be used on all machines.

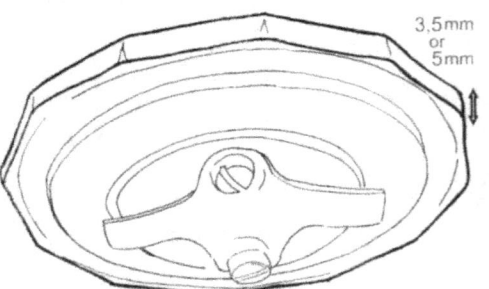

Fuel tank cap 1301 – 7500
Chrome plated brass, material thickness appr. 3,5 mm.

Fuel tank cap 7501– 14015
As for 1301 – 7500, but material thickness appr. 5 mm.

FUEL VALVES AND -TUBES

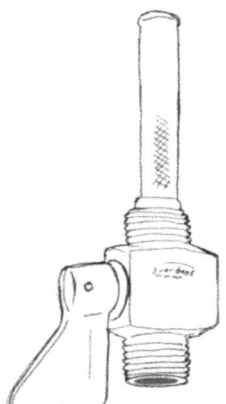

Fuel slide valve 1301 – 1551 Rotating red brass valve, perhaps German make, 'Zöbliz' (MZAG) or German make, 'Everbest'.

Fuel slide valve 1301 – appr. 4500:
Red brass slide valves, German make, 'Zöbliz'.

Fuel valve appr. 4500 – appr. 8500
Rotating brass valve, German make, 'D.R.G.M.-Frankfurt version'.

Fuel valve appr. 8500 – appr. 9000
Chrome plated brass slide valve, British make, 'Ewarts'.

Fuel valve appr. 9000 - 14015
As for appr. 4500 – appr. 8500, but material aluminium.

Fuel pipe 1301 – 1550
Nickel plated brass pipe.

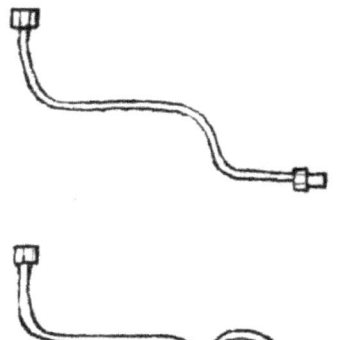

Fuel pipe 1551 – 2050
As for 1301 – 1550, but provided with a pig tail to help reduce the effects of vibrations and tensions, which otherwise can cause the pipe to snap.

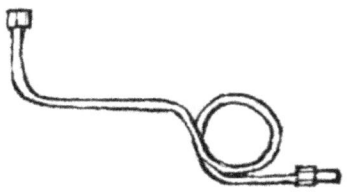

Fuel pipe 2051 – 3863
Pigtailed, but shorter, as the fuel valve was moved from the front to the mid of the tank base.

Fuel hoses with a Union nut, which fits onto carburettor '34, has been a spare part since 1939.

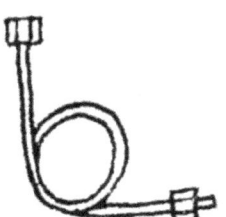

Fuel hose 3864 - 14014
Fuel hose with two identical Union nuts, one at each end of the fuel hose.

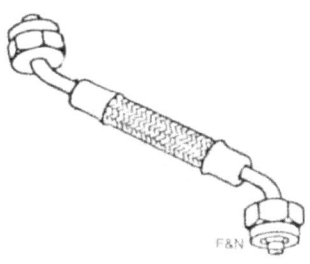

TRANSFERS (DECALS)

All original Nimbus-C transfers are varnished fixed type. There is one transfer on each side of the tank and nowhere else! Water slide transfers and stickers are later imitations.

Transfer 'Standard' 1301 – appr. 5000 (1934 – appr. 1945)
Diameter 86 mm, black back ground with slanted text as well as halo image in gold.

Transfer 'Sport' (1937 – 1947)
Length 96 mm, diameter halo 50 mm. Red back ground, gold text and halo, black outline and shadow.

Transfer 'Special' (1939 – 1953)
Diameter 56 mm. Red background, gold text and halo, black outline.

Transfer 'Standard' appr. 5000 – 14015 (appr. 1945 – 59)
Diameter 56 mm. Black background, gold text and halo.

Enamel badge 12201 - 14015 (1954 – 59)
Pressed brass plate and enameled. White text on a blue background.

'Standard' machines for postal and military services were still provided with the varnished fixed transfer with a diameter of 56 mm.

There are also slide (water based) transfers that look like the enamel badge. There are two versions, one with a golden 'halo' and white text on a blue background, and one with a golden 'halo' and white text on a red background. It is unknown if these transfers were applied by the factory.

ENGINE

The Nimbus-C engine consists of the following main parts:
- Cylinder block with lubricating system
- Cylinder head with poppet valves
- Crankshaft with piston rods, pistons, flywheel and clutch
- Camshaft housing with camshaft and rockers
- Oil sump with kick starter system
- Exhaust system
- Dynamo and electrical system

The carburettor and gearbox are described separately.

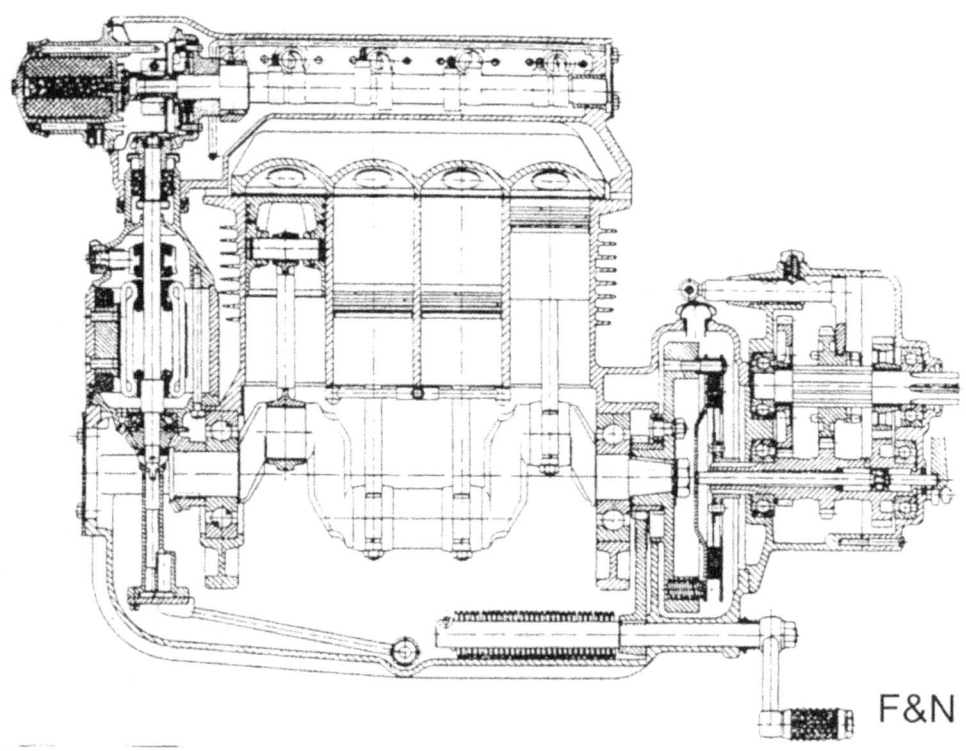

CYLINDER BLOCK

The cylinder block of the Nimbus-C is made of cast iron and is in one piece. The exterior is matt black and the non-critical parts of the interior are painted with red oil repellent enamel.

The casting is processed by milling and boring. At first sight, all cylinder blocks look the same, but there are a few versions. Most prominent are the vertical cooling fins, which cross the horizontal ones at the front and the rear on the oldest cylinder blocks. The cooling fins of the newer cylinder blocks are open at the front and rear side.

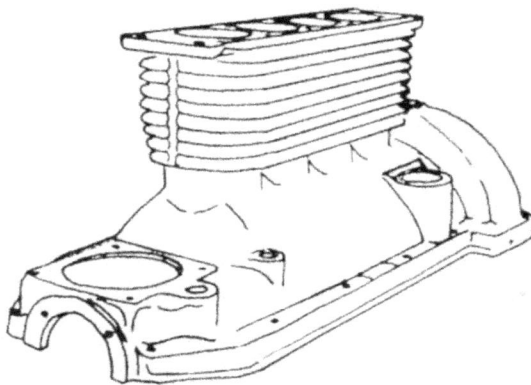

Cylinder block 1301 – 1550

The lube oil pipe to the gear box runs inside the cylinder block on the exhaust side. Therefore, there was a recess for this pipe between the main bearing and the block wall. Because of lubrication problems, most (but not all) machines from 1301 - 1550 were recalled to change the lube oil system. The lube oil pipe was moved to the opposite side, and holes for the lube oil nozzles were drilled. The now unused reces at the right hand side were closed by a dummy plug.

On these cylinder blocks, a clear casting joint flash can be seen on each cylinder, a few centimeters below the bottom cooling fin. There is no clear raised section of metal for the engine number at the left hand side. The edges of the cylinders on the side of the oil filling hole are milled through and the horizontal cooling fins are, as mentioned before, crossed by two vertical ones.

Cylinder block 1551 – appr. 2400

The lubricating system was changed from 1551. Therefore the core of the casting mould was slightly changed too. Besides the changes of the lube oil pipe and nozzles, which was carried out on some of the first blocks (see 1310 – 1550), a flat elevated section in the material of the joint flange below the carburetor was introduced to stamp the engine number onto. Apart from these adaptations, the cylinder blocks remained unchanged, and are the same as those from 1301 – 1550.

Cylinder block appr. 2400 – appr. 6500
The exterior of the design was as 1301 – 2400, apart from the way of milling near the oil filling hole, which resulted in a circular cut instead of a straight cut.
The raised section of metal was half way along the gasket flange below the carburettor, but some engine blocks had this raised section more towards the dipstick.

During a period around 1945 – 47, cylinder blocks were probably cast to stock. It is therefore possible to find castings without vertical casting fins at the front and the rear side (new casting molds) as low as numbers around 6500. By the same token, there are cylinder blocks produced with the old casting mould as high as number (appr.) 7500

Cylinder block appr. 6500 – 14015

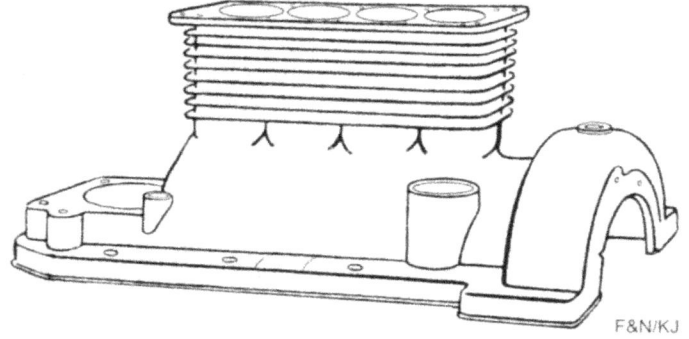

The inside is as before, but the outside had a different cooling fin pattern, whereby the cooling fins run uninterrupted around the front and the rear cylinder. Apart from that, these cylinder blocks are said to be casted with an inferior and less wear-resisting material than before. In addition to the series that ended with 14015, the factory supplied a couple of hundred extra six-digit blocks, starting with no. 100001.

Cylinder block 14171 – 14198
In 2015 the Nimbus club 'Danmarks Nimbus Touring' had cast a series of cylinder blocks with a mould that was manufactured in 1989.
The casting mould meets the same specifications as those at the factory in the fifties.

Threaded plug 1301 – 2560
The hole in the top of the flywheel housing of the cylinder block is closed with an 18 x 1.5 mm threaded plug.

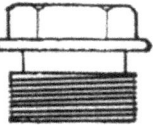

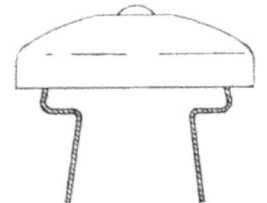

Cap 2561 – 14015
In order to obtain a better ventilation of the flywheel housing, an open cap was fitted instead of the (tightly closed) threaded plug. Although the cap is kept in place with a snap fit mechanism, the 18 x 1.5 mm thread remains in later cylinder blocks, because it is used to secure the engine in the original engine overhaul fixture.

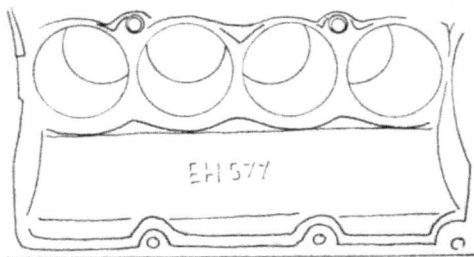

Bore
Cylinders in a factory-new block have a diameter of 60.0 mm. When the cylinders are bored to an oversize (60.6 – 61.2 mm), the new dimension may be stamped in the foremost left hand flange of the dynamo. An oversize of 62.2 mm was not in accordance with the factory specifications, but has been executed over some years without problems.
For military machines, the cylinder diameter is indicated on a special brass plate, riveted to the flywheel housing. A cylinder block can be bored out and fitted with liners. In which case they restarted with 60.0 mm.
The spare parts catalogue mentions a first piston oversize of 60.5 mm. It is unknown if this was ever used

LUBE OIL SYSTEM

The oil system of the Nimbus-C consists of a lube oil pump with a gun metal housing, steel gear wheels and copper lube oil pipes with brass lugs.

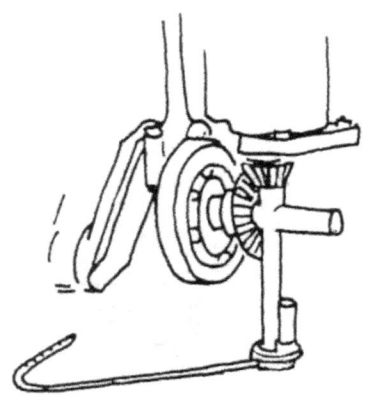

Lube oil system 1301 – 1550
The earliest lube oil system is a pressure system, where oil under pressure flows to all places where it is needed. The oil for lubricating the crankshaft flows through a spring loaded copper duct at the front to a bore in the crankshaft. This bore is connected to each of the main journals. From there, the oil flows through the bores in the piston rods to lubricate the piston pins and pistons (See crankshaft). A second oil flow runs from the cylinder block to the lay shaft in the gearbox, from where the oil flows back to the oil pan (See gearbox).

A round cover on the front of the engine gives access to the spring loaded pressure relief valve of the oil pump, which, only when the valve is open, supplies the oil to the gear box. Another lube oil pipe runs to the cam shaft housing, where the oil passes through the pressure-operated generator current cut out relay and flows further to the camshaft bush bearings. From the rear bush bearing, the oil flows over four ribs (weirs) on the bottom of the cam shaft housing, forming small oil pans, where the cams are lubricated by dipping. The oil flows along the dynamo gear wheels back to the oil sump through an oil return pipe (See camshaft housing).

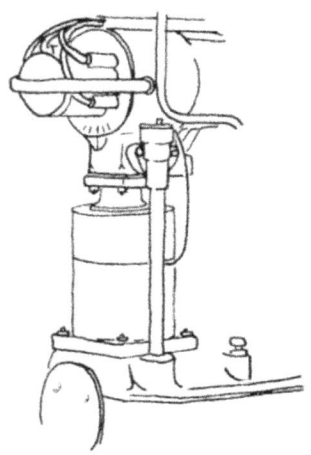

The oil pressure-operated generator current cut-out relay in the lube oil pipe to the cam shaft housing is part of the electrical system. The relay is connected between the dynamo and the battery and prevents current flowing from the battery back through the dynamo when the engine stops and oil pressure falls back. Only when the engine is running is there oil pressure, and also a demand for electric current for charging the battery. A suction pipe runs from the lube oil pump to the bottom of the oil sump, where an oil strainer, in combination with the cover, filters the oil.

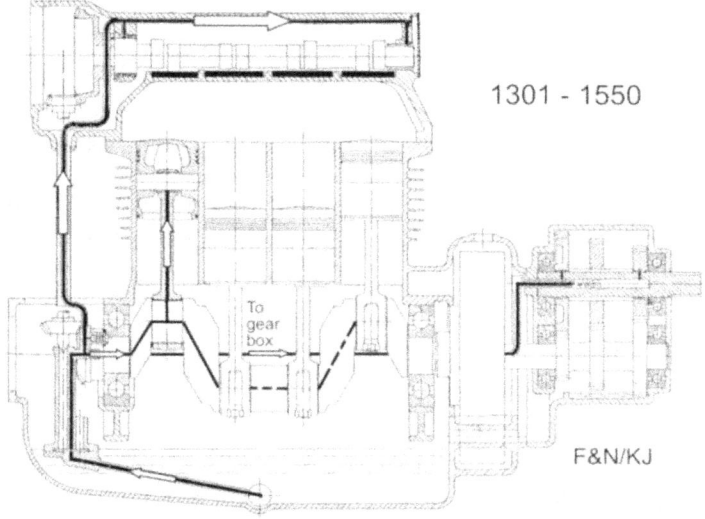

1301 - 1550

F&N/KJ

Lube oil system 1551 – 14015

Because of problems, due to insufficient lubrication and clogging of the oil ducts in the crankshaft, the lube oil system was modified from a pressure- to a drip and splash system for lubricating Crankshaft and pistons. The lube oil pump was modified such that the oil flow was split and oil was directed to two different lube oil pipes, one to the camshaft housing and another to nozzles in the crankcase as well as directly to the gearbox (and no longer through an open pressure release valve). At the same time, the dynamo was changed (see dynamo).

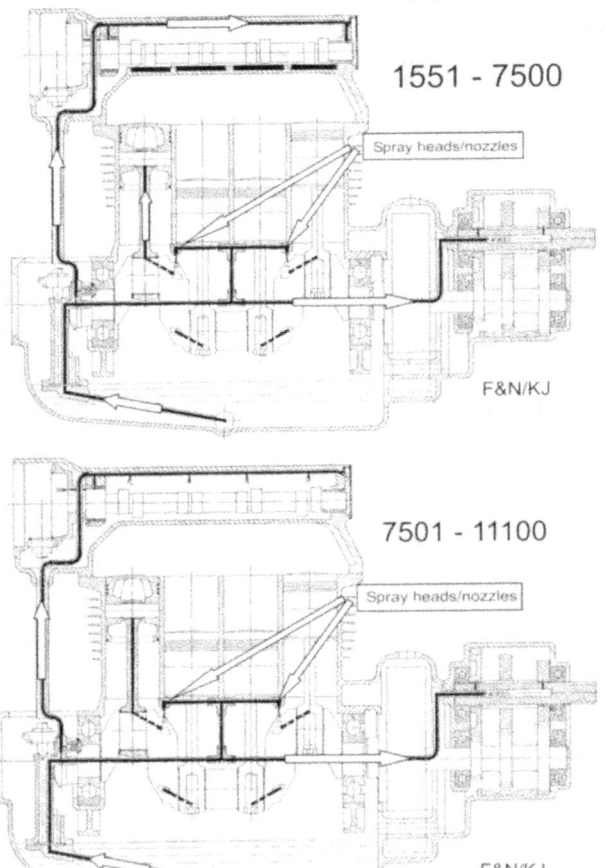

A voltage- and cut-out regulator was fitted, and the oil pressure operated current cut-out relay in the lube oil line to the cam shaft housing became obsolete.

The lube oil system for the camshaft housing (see chapter 'Camshaft housing') was slightly modified from machine 7500 (see chapter 'Camshaft housing'). Instead of lubricating the cams by being dipped in oil puddles, formed between the ribs at the bottom of the camshaft housing, the cams and gear wheels were now lubricated by means of holes in the lube oil pipe.

The second lube oil pipe previously mentioned above provides both the gear box and the nozzles in the crank case with lube oil. The pipe runs along the left-hand inside of the cylinder block and splits up half way at a T-piece, which again splits up the oil flow - to the gear box on the one hand and to each of the two nozzles on the other, the latter fitted in two slots in the block.

Some older lube oil pipes have extended nozzles. These have been used in 1934 engines after being adapted to the new lubricating system

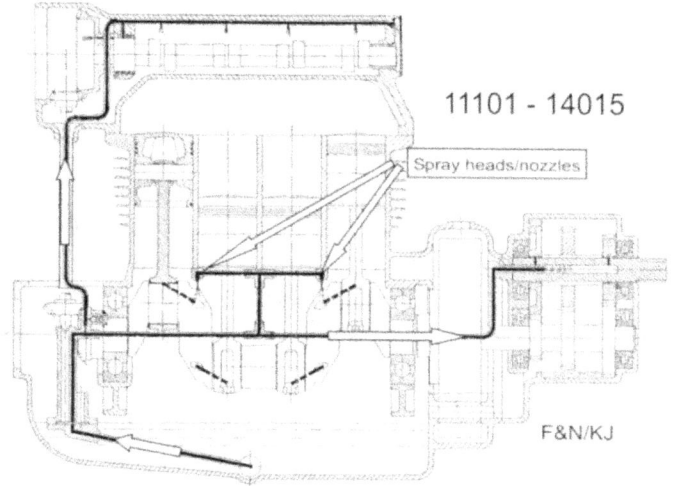

*Lube oil return
pipe 1301 – 1550*

The lube oil return pipe is from matt nickel plated brass. The lube oil pressure pipe is made from copper and has a pressure operated current cut-out relay.

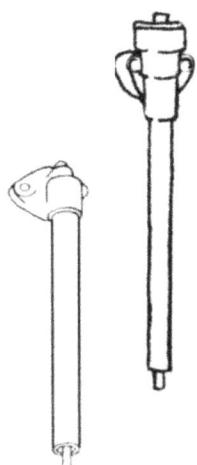

Lube oil return pipe 1551 – 14015

The oil pressure pipe to the camshaft housing is fitted coaxially inside the lube oil return pipe.

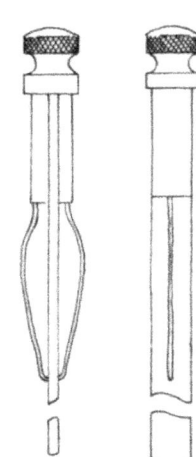

Sounding rod (dipstick)

The sounding rod or dipstick, as we call it today, comes in two versions:
1301 – 1550: with a handmade small head.
1551 – 14015: with a machined, larger head.

CYLINDER HEAD

The cylinder head of the Nimbus-C has basically been unchanged during the whole production period; there are however two versions: one is flat, using a plain flat head gasket, the other is spiggoted and utilises gasket rings.
The cylinder head with cooling fins and intake manifold is cast as a unit in grey cast iron. The valve guides are made from cast iron and are pressed in holes into the cylinder head. The cylinder head is matt black, like the cylinder block.

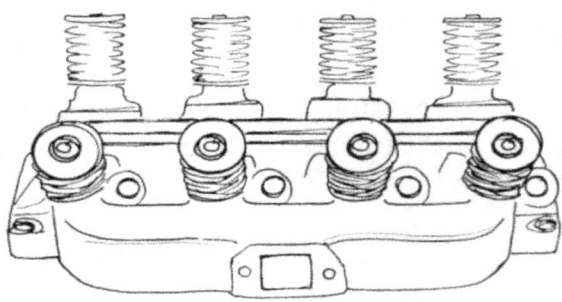

Cylinder head 1301 – 1550
The main joint of the cylinder head is flat and designed to be used with a flat asbestos-copper gasket. The bottom edge of the intake manifold is sharp and the bore of the intake manifold is square.

Cylinder head 1551 – 8500 As 1301 – 1550, but with an oval bore in the intake manifold.

Cylinder head 8501 – 13572

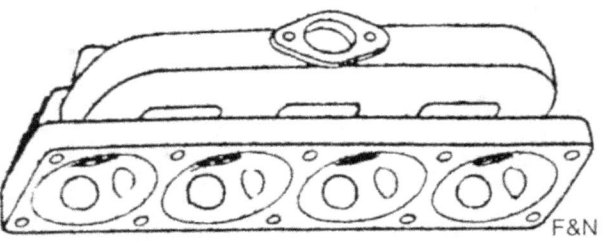

There are milled recesses in the base of the cylinder head around the combustion chambers to accept special gasket rings. These gaskets, also called gas rings, will result in better cooling of the cylinder head as compared with the isolating flat copper-asbestos gasket, because the cylinder head and –block can be tightly fitted together. The bottom side of the intake manifold is shaped the same way as the top side, but with a couple of casting stubs.

Some cylinder heads have the circular recesses removed by milling down the base and hence can be used with a flat gasket like the early version of the cylinder head.

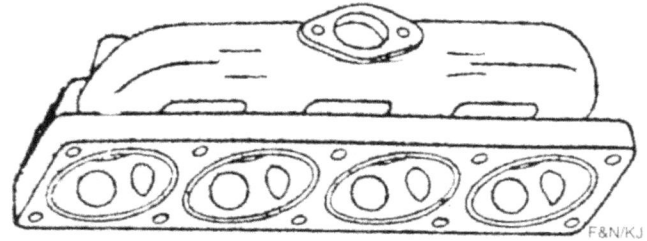

Cylinder head 13573 – 14015 (not all)
Same as 8501 – 13572, but a 4 mm hole is drilled in the material surrounding the valve guides to accept a guiding pin. This pin is fitted for the closed aluminium valve housing around valve and valve spring.

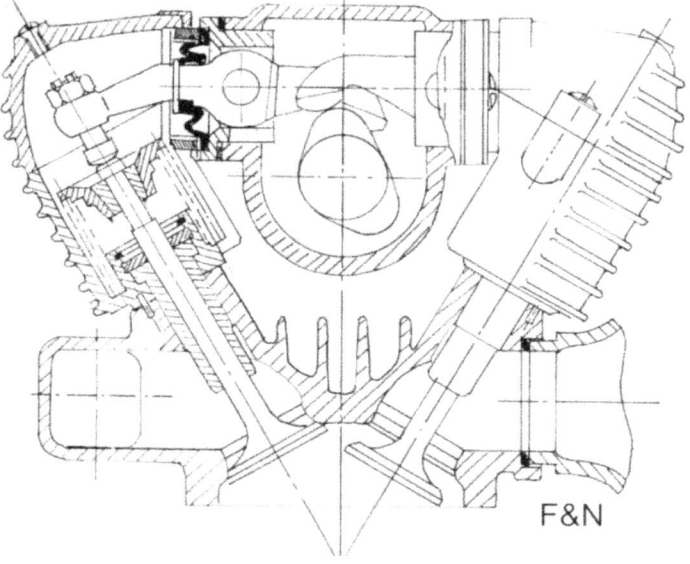

»Police cylinder head«
A very few cylinder heads have been supplied by the factory to the traffic police with a reduced combustion chamber and hence a higher compression and increased power.

Threaded hole in inlet manifold
Some cylinder heads have a threaded hole in the inlet manifold, closed-off with a plug. This hole can be used to connect a vacuum meter, to test the engine. The machines for the army especially were tested in this way.

VALVES

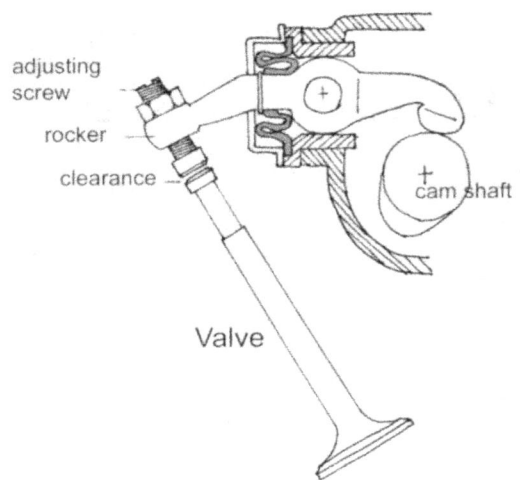

At first sight, all valves appear to be the same. In the course of time however, different metal alloys have been used, and in many cases, intake- and exhaust valves were made from different materials. Sometimes they were marked on the valve heads »NIMBUS-INT« on the intake- and »NIMBUS-EXH« or just »EXH« on the exhaust valves.

In other instances, intake- as well as exhaust valves were marked »NIMBUS-S« near the end of the valve stem.

Part, but not all, of the exhaust valves can be identified by the fact that they are not attracted by a magnet.

VALVE GUIDES

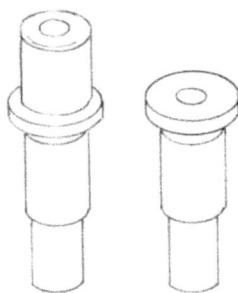

Valve guides 1301 – 13572
Designed to accept *two* valve springs between the upper and lower spring cups.

Valve guides 13573 – 14015
Designed to accept *one* valve spring in the valve housing.

VALVE SPRINGS

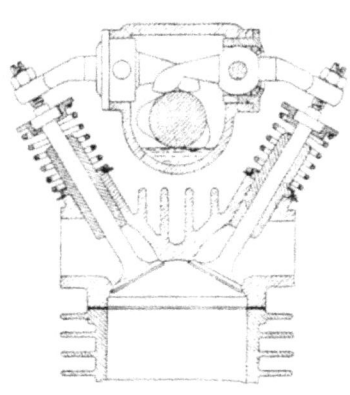

Valve springs 1301 – 13572
Two valve springs: the outer is left hand wind with 6 turns and a wire thickness of 3.0 mm, the inner is right hand wind with 7 turns and a wire thickness of 2.0 mm.

Valve springs 135473 – 14015
One valve spring, stronger, left hand wind with 5 turns, wire thickness 3.25 mm, designed for Nimbus-C with enclosed valve housings.

SPRING CUPS

Upper valve cup 1301 – 2050 The valve is secured by means of a horse shoe shaped cotter, fitted in corresponding slot in the upper spring cup. The upper spring cup is matt nickel plated steel, later cadmium plated steel were applied.

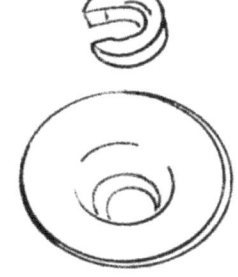

Upper valve cup 2051 – 14015 The cotter was modified and consists of two halves. This made it possible to make the valve stem thicker in this area. The shape of the upper valve cup also changed to accept the new cotter.

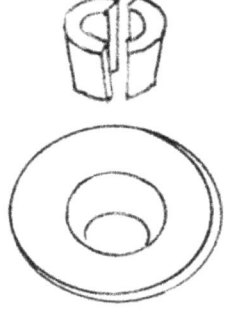

Lower valve cup 1301 – 13572 The lower valve cup rests upon a mica washer, to isolate it from the cylinder head.
The material for the lower valve cup is either hardened special steel or cadmium plated steel.

Valve housing 13573 – 14015 The valve spring rests upon the bottom of the aluminium valve enclosure housing which is isolated from the valve guide by means of an asbestos disc at the side of the exhaust manifold and a rubber disc at the side of the inlet manifold. The valve enclosure housings can in fact be fitted onto all earlier machines. The factory issued an instruction leaflet with the necessary changes to be made (requirements of the rockers and valve guides in the cylinder head) and required spare parts.

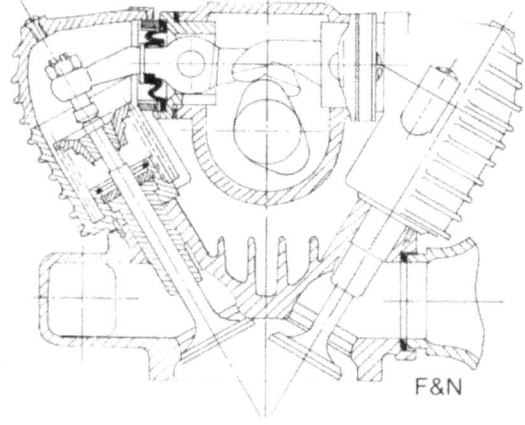

An entirely smooth valve enclosure, supplied by Isidor Meyer, Copenhagen, hit the market a few years earlier as a non-original auxiliary device.

CRANKSHAFT

The crankshaft of the Nimbus-C is drop forged out of one piece. It rotates in two main (ball)bearings. The big end bearings for the connecting rods (conrods) are standard 40 mm in diameter. The big end journals can be ground when they have reached the critical limit for ovality. According to information from the factory, the specifications for grinding are 39.5 and 39.0 mm. Later on, journals were ground down in steps of 0.25 mm, if possible. Where newly fabricated connecting rods are used, it is possible to grind the journals down to 38.0 mm. When a crankshaft is ground, all journals must be ground to the same dimension (See chapter 'Numbers etc'). There were two main lubricating principles for Nimbus-C, described below.

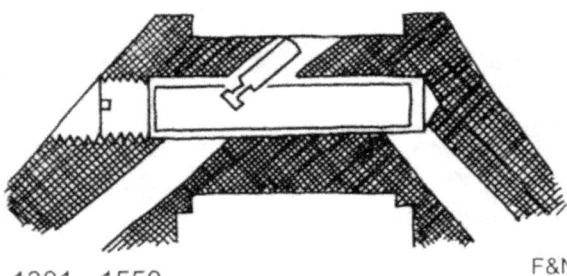

1301 - 1550 F&N

Crankshaft 1301 – 1550

The first version of crankcase lubrication was lubrication by pressure. The oil flows from a connection on the lube oil pump through a bore in the crankshaft to each of the four rod journals. To divide the oil over the journals to lubricate each big end bearing, each journal has a slotted tube, which holds a small restricting pin in place in the bore of the journal. This is a brilliant system, but the lube oil at that time did not have 'cleansing' properties, and had the tendency to build sludge and cause the oil bores to be clogged, leading to a lack of lubrication and damage to the engine.

Crankshaft 1551 – 7500

The next lube oil system is a drip and splash system. Because of the risk of oil ducts becoming clogged, the pressure system was converted to a drip and splash system. The big ends are now being lubricated by an oil pocket in the crankshaft, which collects the oil from the two nozzles. As a result of the centrifugal force, oil is forced through bores
('Cines eyes') to the big end bearings. The owners of machines 1301 – 1550 were offered a modification of the lube oil system of their motorcycle, free of charge. It is unknown how many owners accepted this offer.

Crankshaft 7501 – 11100
As 1551 – 7500, but the oil pocket was changed to a simple slanted bore, the far part nicknamed 'Chinese eye'. When the crankshaft is reground, this bore may give problems, because, due to the reduction of the crank diameter, these bores are no longer half way along the big end bearing, but are shifted to one side. The more material is ground away, the more asymmetric the distance becomes to the edges of the journals, leading to poor lubrication at one side.

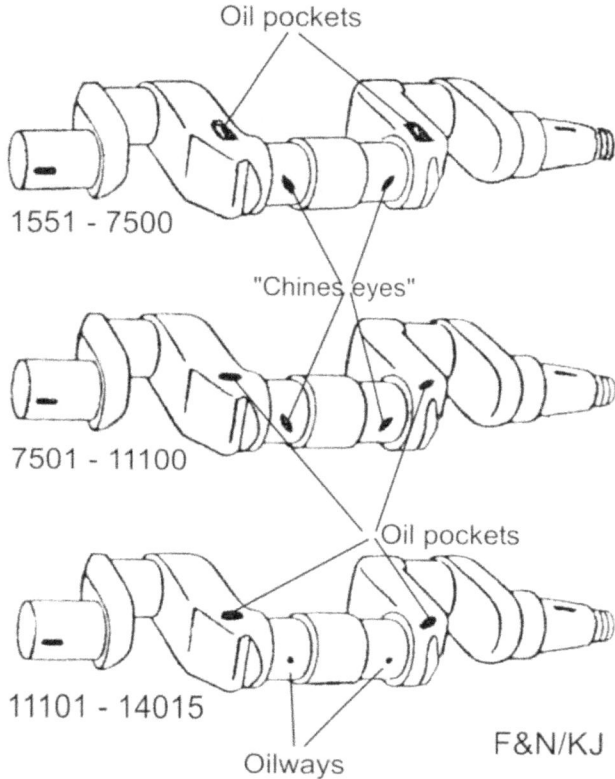

Crankshaft 11101 – 14015
As 7501 – 11100, but the slanted bore no longer runs straight to the rod journal, but makes an angle. The far part of the oil bore (oilway) is drilled into the rod journal at a right angle. This construction allows the rod journal to be ground, but the oil bore will remain in the centre of the big end bearing, so an even lubrication is maintained.

CONNECTING RODS

The connecting rods of Nimbus-C are drop forged, with a centrifugally cast white metal bearing for the big end, and a pressed-in bronze bush bearing for the small end.

Connecting rods come in two versions, one with a hollow shaft, the other with an I-shaped profile. The rod with the I-shaped profile comes again in two versions. The main difference between the latter versions is the bottom half of the big end bearing clamp, which is different in thickness and weight. The crankshaft must be fitted with bottom half bearing clamps having the same thickness and weight.

The conrod shaft is generally marked at the side, opposite to the lube oil pipe, with the corresponding cylinder number. In addition, the base of the conrod and the corresponding lower half of the big end bearing are marked to match, but differently from the

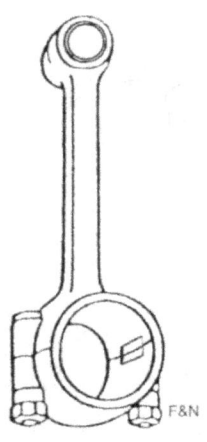

other pairs, to prevent wrong orientation or inter changing. This can be any marking.

Last but not least, the conrods can be marked with a colored band around the shaft as identification for the dimension (see chapter 'Numbers etc.')

Connecting rod 1301 – 11100

The conrods have become a part of the engine's lube oil system; the white metal bearing of the big end is drilled through to meet the hollow shaft of the conrod, which is connected to the oil groove in the small end bush bearing.

Connecting rod 1301 -7500

The conrod and the bottom half of the big end bearing are clamped together with bolts with a hole for a split pin, to lock a castellated nut. After 7500, locking was done with a spring washer, so the hole was left out and non-castellated nuts were used.

Connecting rod 11101 – 14015

The shaft of the conrod is I-shaped, does not have an oil bore and has the 'NIMBUS' mark cast into it. The small end bush bearing is thinner and no longer has an oil groove but an oil hole on top instead. The bottom bearing half has a robust reinforcing bulge. As mentioned before, these conrods come in two different versions with different weight, which may not be combined.

Newly produced conrods of this version may be marked with the name "Lillehøj" cast-in.

PISTONS

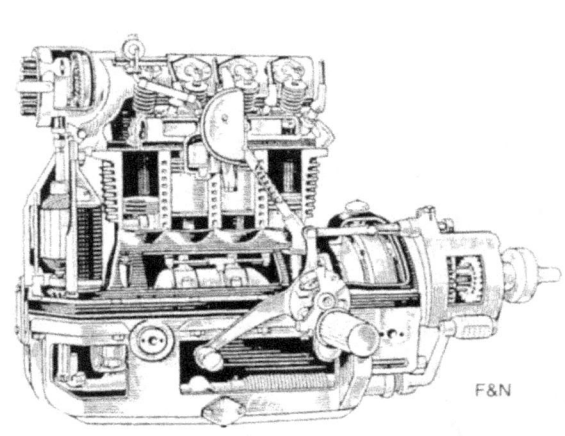

Normally, Nimbus-C pistons are cast in aluminum. Cast iron pistons with an oversize to 60.6 and 61.2 mm have been found, but it is unknown if they were bought in by the factory. In the course of time, the factory has used many aluminum versions, sometimes during the same period. The information below may therefore not be fully complete.

Piston 1301 – 2400

Flat top, three piston rings above the gudgeon (or wrist) pin and one below. The gudgeon pin is retained in place with brass plugs at each end that are not secured.

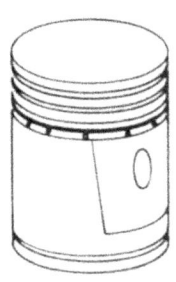

Piston 2401 – 2859 (5000 for 'Standard' and 'Luxus')

As 1301 – 2400, but without the brass plugs in gudgeon pins. Locking was done with circlips in a groove at both ends of the gudgeon pin bore instead.

Piston 2860- 3625 ('Sport')

Round top, remaining features as 2401 – 2859. This version is still produced today (2016)

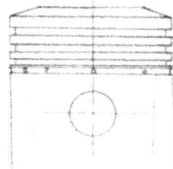

Piston 3626 – 5000 ('Sport' and 'Special')

As 2860 – 3625, but because the compression was too high, and the changed (poorer) quality of the fuel at that time, the top of the piston was milled down slightly.

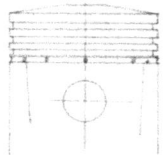

Piston 5001 – 9000

As 3626 – 5000, but the piston was produced such that the final milling process (required for 3626 - 5000) was no longer necessary.

Piston 9001 – 11100

Crown as 5001 – 9000, but machined oval (marked "oval" on top). The number of piston rings was reduced to just three above the gudgeon pin. The clearance was reduced from 0.08 or 0.09 mm to 0.07 mm, measured at the largest diameter, at the lower end of the piston skirt. Ovality is 0.12 mm. The relieved bearing areas around both ends of the piston pins were no longer employed. (As seen on the 3 pistons drawn above.)

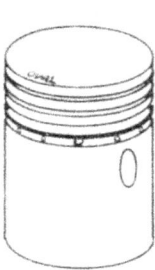

Piston 11101 – 14015

Flat top, but the rest as piston 9001 – 11100. The third ring from the top, the scraper ring, was replaced by a wider, more modern design of oil scraper ring (See chapter "Piston rings").

Dealers at that time are said to have been unsatisfied with the current pistons. Therefore some of them fitted other pistons with four rings instead, either all four up or three up and one down.

PISTON RINGS

All original special steel piston rings for Nimbus-C are 2.5 mm wide and are 2.4 mm thick. The dedicated oil scraper rings can have different dimensions.

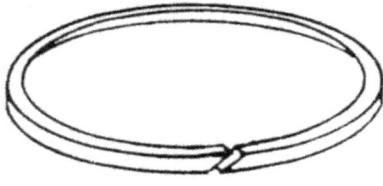

Piston ring 1301 – 4500 (5000 for 'Standard' and 'Luxus')
All piston rings have the same slanted gap. Rings are fitted to the piston with their gaps sloping alternately right and left.

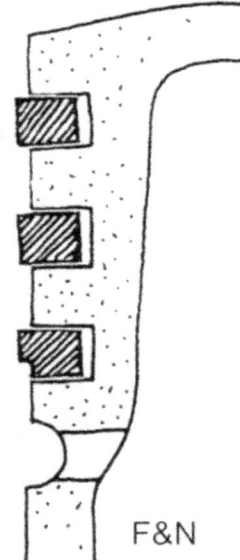

Piston ring 4501 – 5000 (3rd ring)
The third ring from the top machined down to create a smaller contact surface, resulting in a higher pressure on the cylinder wall. As it is the lower edge that is milled down, the piston ring will act as an oil scraper ring.

Piston ring 5001 – 14015 (9000 for 'Standard' and 'Luxus')
All piston ring gaps changed from a slanted gap to a vertical gap.

Piston ring 9001 – 14015 (1st ring)
The upper piston ring is chrome plated. The ring is slightly tapered in section and is therefore marked 'TOP' on the surface that must be up when fitting. Apparently this tapered ring was intended to reduce wear of piston ring and cylinder wall.

Oil scraper ring 11101 – 14015 (3rd ring)
This third ring may be constructed as a 4 mm thick oil scraper ring. It can be of different makers. Some of these rings are fitted with an expander to assure sufficient pressure onto the cylinder wall. One of the expander versions is marked »C35«; this mark may be visible through the piston ring slot when fitted.

FLYWHEEL

The flywheel of the Nimbus-C comes in three versions, all with the same diameter (See kick starter and clutch).

Flywheel 1301 – 8000
Weight 5270 g. Milled down such, that the bolts of the clutch plates can be secured with cotter pins or wire.

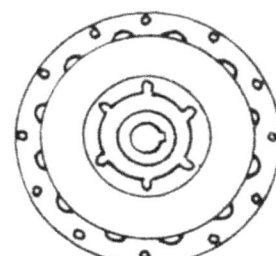

Flywheel 8001 – 9000
Weight 3740 g. The weight reduction was achieved by milling down the material. Otherwise as 1301 – 8000.
Some flywheels were milled down even further to reduce weight, e.g. for racing. It resulted into a better acceleration but idle running was unstable.

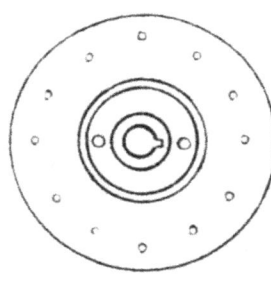

Flywheel 9001 - 14015
Weight 4200 g. No milled down profile for securing clutch bolts. The construction of the kick starter unit was changed (see kick starter).

Flywheel 1301 – 9000 can be converted such that it will fit kick starter system 9001 – 14015.

KICK STARTER

There are two constructions of the kick starter system for Nimbus-C:

Kick starter system 1301 – 9000 In the flywheel, a keyway for a special steel pawl disc, which is fitted on top of a conical compression spring, is made. A combination of the gear wheel and the pawl wheel makes the flywheel rotate, when the kick starter is trotted down. Both, pawl disc and pawl wheel come in different ver-

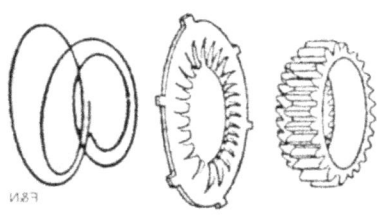

sions and different dimensions. The construction is very unstable when worn.

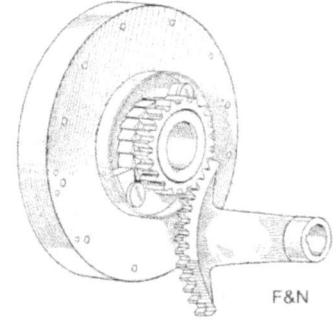

Kick starter system 9001 – 14015 The flywheel is fitted with two springloaded pawls, which are constantly engaged into the kick starter tooth wheel. It forms a stable construction, which seldom leads to problems.

The kick starter system 9001 – 14015 can be fitted onto flywheels 1301 – 9000.

CLUTCH

The clutch of the Nimbus-C is a dry, single plate clutch with steel pressure- and counter pressure plates. It consists of a clutch plate with hub (and drivers), 12 springs, friction lining, steel bushes and -bolts.

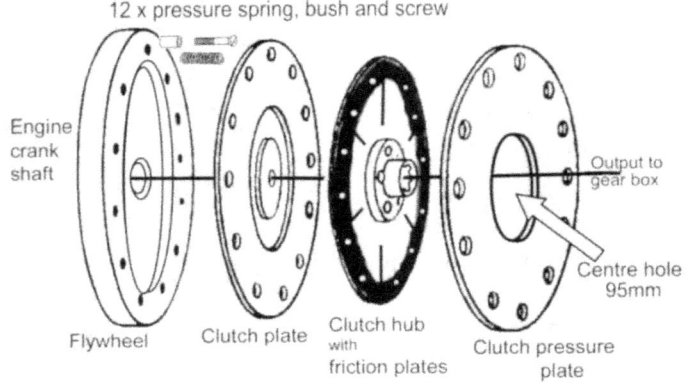

Clutch 1301 -1550
Pressure- and counter pressure plates without lining. Plate thickness was at some point of time before 1550 changed from 2 mm to 2.5 mm. Central hole in counter pressure plate 95 mm. The 2.5 mm thick clutch plate has 6 straight slids and was lined with riveted-on friction material.

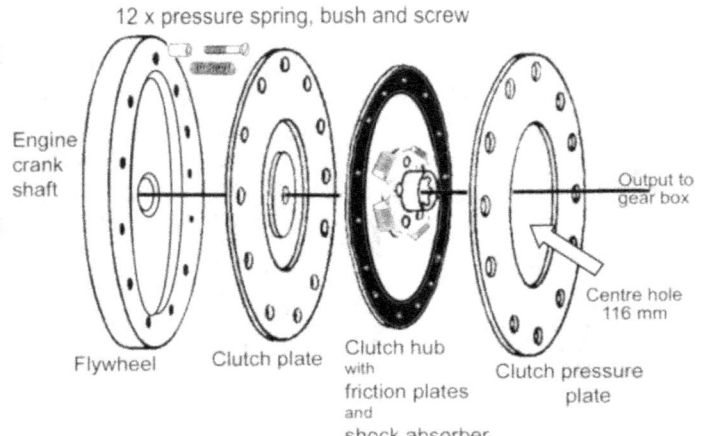

Clutch 1551 – 1984
As 1301 – 1550, but the central hole in the counter pressure plate was increased to 116 mm to create room for the shock absorbing springs in the clutch plate with hub.

Clutch 1985 – 2400
As 1551 – 1984, but with hardened stoppers in the shock absorbing springs.

Clutch 2401 – 2560
Clutch plate as 1985 – 2400 but without lining, with 6 straight slids and shock absorbing springs with hardened stoppers. Pressure- and counter pressure plates with riveted lining.

Clutch 2561 – 4500 Clutch plate without lining with 6 straight slids. Pressure- and counter pressure plates with riveted-on lining.

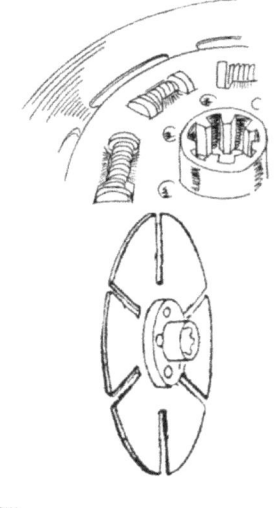

Clutch 4501 – 9000
As clutch 2561 – 4500, but with rounded as opposed to straight ends of the slids, to prevent tearing in the corners of the slids. Lining is either vulcanized or riveted with countersunk heads.

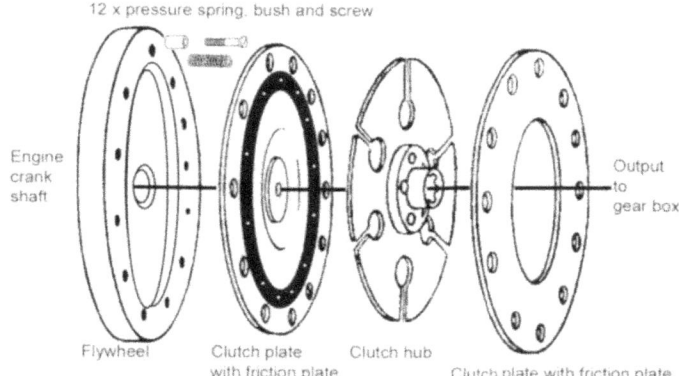

Clutch 9001 – 14015
As the flywheel was modified, the clutch followed. The clutch plates consist of a hub with steel riveted-on spokes and with linings at both sides. Diameter 159 mm. Pressure and counter pressure plates are without lining.

To-day design 9001 – 14015
For flywheel from 9001, a new and better clutch was produced with oil safe lining and otherwise identical to clutch 4501 – 9000.

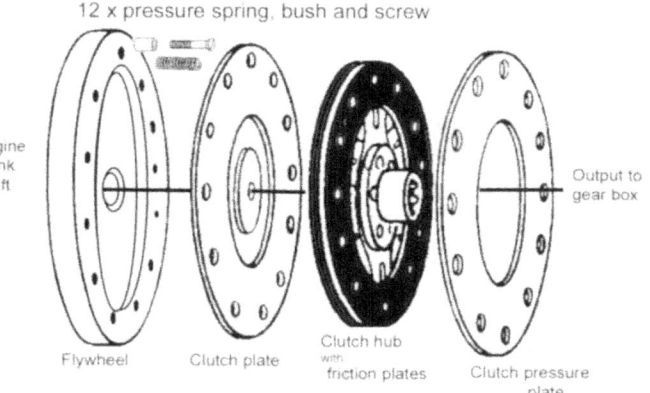

CAMSHAFT

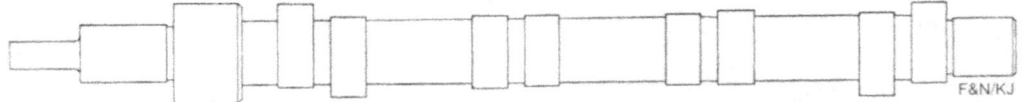

The camshaft for Nimbus-C is drop forged for all production machines. Experiments have been carried out with specially made camshafts having higher lift and different profiles for valve control, particularly for racing and product development testing.

The camshaft rotates in phosphor-bronze bushes fitted in the aluminium camshaft housing. In the sides of this housing, the rockers are pivoted. At the front end of the camshaft, a gear wheel is fitted that is driven by the dynamo's top gear wheel.

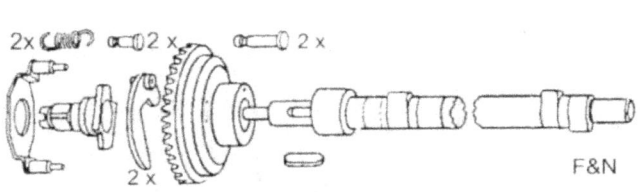

A pair of spring loaded rotating weights are fitted to the cam gear wheel. They drive the mechanism for regulating the ignition timing. At the front, in the camshaft case, we find the distributor housing with contact breaker, condenser, with the distributor and ignition coil as a unit.

Camshaft gear wheel 1301 – 3000

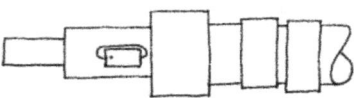

Gear wheel made from special steel and has skew cut teeth, with two holes for the ignition driving pins and two for the lock plate. In order to advance the camshaft in relation to the crankshaft, the camshaft gear wheel was given an offset by using an offset key for production numbers 2551 – 3000. Both camshaft and gearwheel remained unchanged.

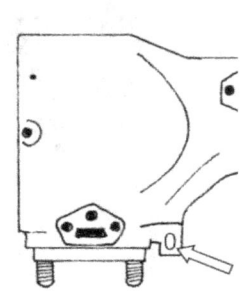

It goes without saying that this offset key may also be used for machines 1301 – 2550. The use of the offset key is indicated by a "0", stamped onto the camshaft housing front left hand foot (near the dynamo).

Camshaft gearwheel 3001 – appr. 10000

The position of the keyway is now such, that a normal key can be used. To distinguish between the camshaft gear wheels 3001 – appr. 10000 and those which need an offset key, the new gearwheel is marked "2". In order to reduce the noise of the skew-cut camshaft gearwheels, tests have been carried out with gearwheels made from the fiber material "Durkonton". Gun metal has also been used. The number of machines and the period during which these camshaft gearwheels were fitted, is unknown.

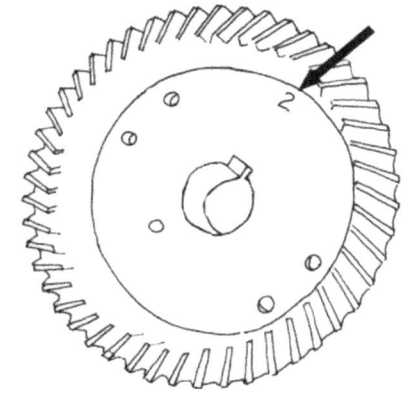

Gear wheel appr. 10000 – 14015

Construction as camshaft gearwheel 3001 – appr. 10000 but with straight cut teeth.

Some of the gearwheels from 1956 – 59 show a star. At some point in time during that period something went wrong in the production plant, mainly because one of the drawings was misinterpreted.

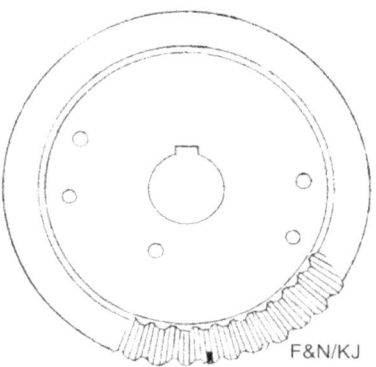

An unknown number of camshaft gear wheels were incorrectly processed; the two holes for the ignition studs and the two for the retaining plate were drilled at wrong positions. This resulted in the ignition having an incorrect dwell time and other problems related to ignition timing after assembly. In the factory the gear wheels were reworked by plugging off the wrongly drilled holes and new holes were drilled at the correct positions.

A gear wheel that was reworked according to this procedure is marked with the number "2" or a star. But be aware that the "2" and the star can have different meanings!

At some point in time, gear wheels with correctly placed holes (ie: did not require re-work) were incorrectly marked with the number "2" or a star. These marks were cancelled with an "X".

Attention should be paid to gear wheels marked with a star or a "2" as, irrespective the rework procedure, an unknown amount of incorrect gear wheels still hit the field.

New construction
All four gearwheels of the timing system are heavily stressed and are subject to wear and tear, especially the upper dynamo gearwheel and the camshaft gearwheel. All four gearwheels have been newly produced since 2004 and are available today with helical cut teeth.

CAMSHAFT HOUSING

The camshaft housing of the Nimbus-C is cast in aluminium. The basic construction has been unchanged during the whole production period, but some modifications were made.

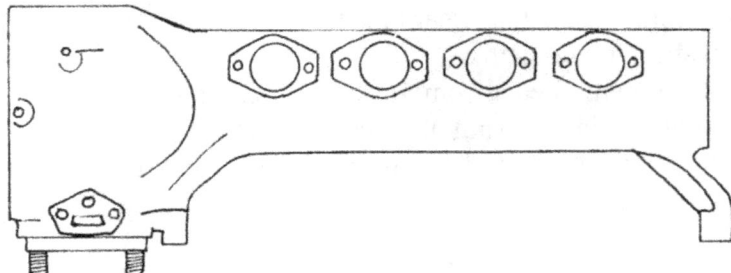

Camshaft housing 1301 – 2050
There are two bearing bushes fitted in the camshaft housing, which get their oil through a lube pipe. These are accessible through two holes at the opposite ends of the housing (see picture and 'C'), closed by two M4 screws. The oil from the rear bearing bush flows over a couple of raised sections on the base of the housing (see picture and 'E'), and then past the two front gearwheels, back down to the oil sump through a rectangular hole. The raised sections are between the cams, and serve to provide an oil reservoir to lubricate each cam lobe. The camshaft and rockers are lubricated by oil-splash from the cam lobes, dipping into the oil in the camshaft housing base. There is a mark at the lower front end of the camshaft housing for the

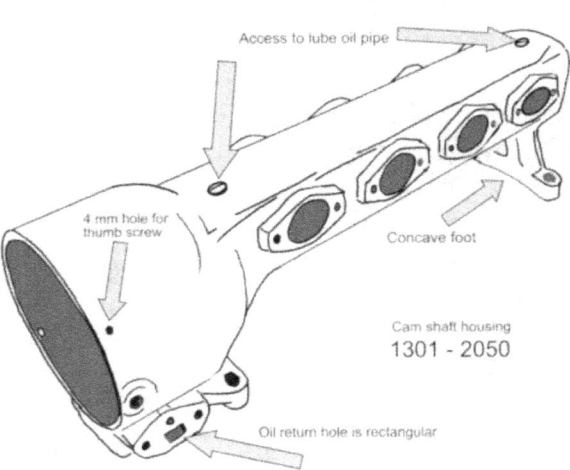

ignition timing in combination with a corresponding scale below, on the ignition coil housing.
The ground connection for the ignition is accomplished by a thumb screw, placed right over the bracket at the left hand side. Please note that the 'feet' of the camshaft housing are concave and that the oil return duct is rectangular.

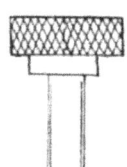

Camshaft housing 2051 - 7500
As 1301 – 2050 but contrary to the original thumb screw as an earth connection, there is now a boss and a screw for the adjustment plate fitted to the scale. (The adjustment mark and scale for the ignition at the front are still there but no longer serve any purpose). The 'feet' are now flat at the base and the oil return duct is trapezium-shaped.

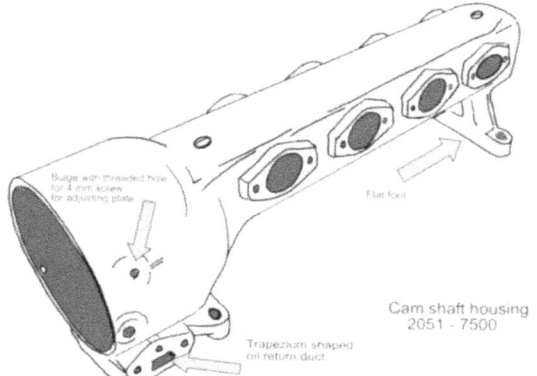

Camshaft housing 7501 – 13572
The lube oil system for the camshaft housing was modified. The lube oil pipe was provided with a number of holes, one to lubricate the gear wheels (B) and one for drip lubrication for each couple of cams (D). The rear bush bearing is lubricated by the oil that is directed to the rear of this bearing. Both bush bearings are lubricated as before. The access screws at the top are replaced by solid brass pins.

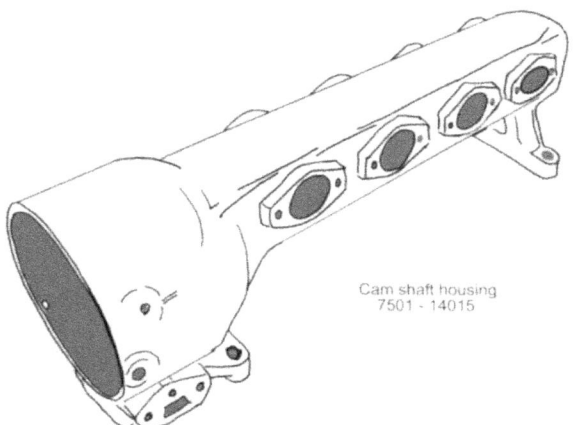

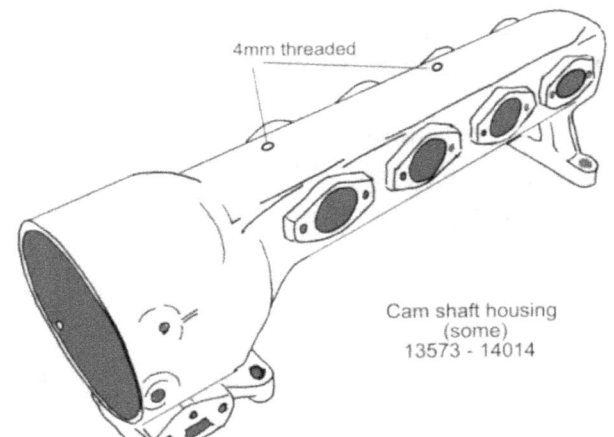

Camshaft housing (13573 – 14015) (not all)
Identical to camshaft housing 7501 – 13572 but because the ignition cables were fitted after fitting the camshaft housing, a couple of 4 mm holes were drilled at the top of the housing near the right hand side.

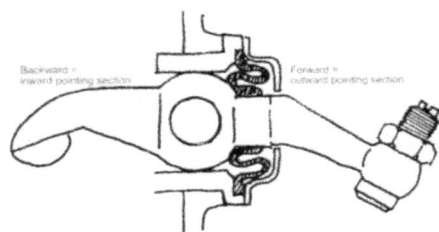

ROCKERS AND ROCKER PIVOTS

Rocker 1301 – 7500
Made from hardened special steel. Material thickness of the backward/inward pointing section is about 5mm.

Rocker 7501 – 13572
As 1301 – 7500, but the material thickness was increased to about 7 mm and the cam follower part is a little bit thicker and more angled.

Rocker 13573 – 14015
As 7501 – 13573, but on the outer side, protruding, there is a lip to hold the rubber bellows in place. The older version can have a soldered ring for the same purpose (see valve enclosure 13573 – 14015).

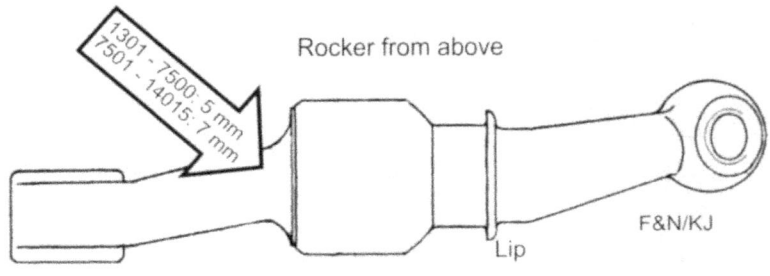

Rocker guide 1301 – 10300
Rocker guide from special steel with a non-removable cover and no gasket. It can be matt nickel plated, later burnished and even later cadmium plated.

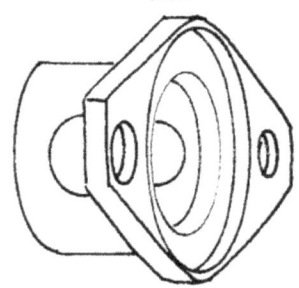

Rocker guide 10301 – 13572 After the high front fork and the open front mudguard was introduced in 1948 (from 7501 on) due to the open front mudguard, the air slipstream while on the move could lead to oil leaking from the rocker. To prevent this, a rubber seal, held in place by the cover was fitted. This gasket has been available since 20014 in a more durable red silicon material.

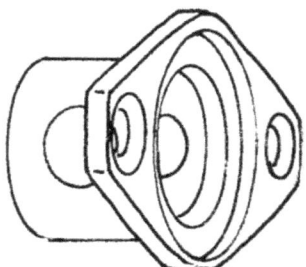

Rocker guide 13573 – 13900
As 13301 – 13572, but with countersunk holes for cheese headed screws. These were needed for fitting the valve and spring finned enclosures. For this rocker guide it was necessary to fit a cover with collar for a rubber ring. The rubber gasket is the same as from 10301.

Rocker guide 13901 – 14015
Because of problems keeping the system with the valve enclosures mechanically tight, the rocker guide was modified and provided with a fixed collar to hold a special rubber gasket.
It is unknown how many of the 115 machines, are fitted with this type of guide.

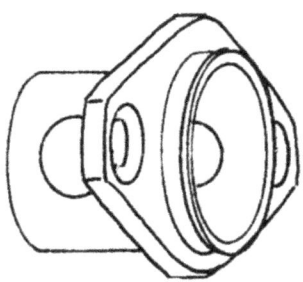

DYNAMO RETAINING RING

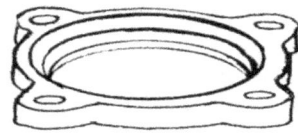

Retaining ring 1301 – 2400
A slender construction made from steel plate; the material and thinness easily lead to deformation when placed under stress.

Retaining ring 2401 – 7500
A robust construction with increased material thickness. Flat base internally.

Retaining ring 7501 – 9000
As construction with 2401 – 7500 but with a conical base internally, in an attempt to achieve a tighter fit.

Retaining ring 9001 – 14015
Construction basically as 2401 – 7500, with the flat base, but with different material dimensions.

OIL PAN (SUMP PAN)

The oil pan of Nimbus-C, which is cast in aluminium, comes in one construction with a few versions:

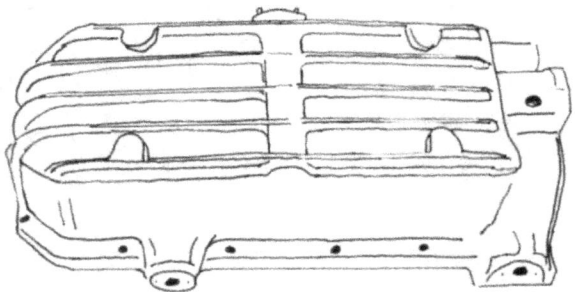

Oil pan 1301 – appr. 1400
On the outside of its base the oil pan has five longitudinal cooling fins; the three middle fins run all the way to the front cover. The outer fins bear four bosses, the purpose of which is unclear. A best guess is that they could be used for fitting a steel protective splash guard onto the oil pan, although threaded holes have never been found. On the inside of the base, at the right hand side, there is a small triangular boss, to guide the lube oil pipe.

Oil pan appr. 1400 – 2400
As 1301 – appr. 1400, but with a reinforced base between the cooling fins under the kick starter.

Oil pan 2401 – 7500
The boss in the oil pan for guiding the lube oil pipe was left out and the bottom ventilation hole in the flywheel housing was placed more forward. This prevents the centre stand covering the hole when the stand is folded up.

Oil pan 7501 – 12000
As 2401 – 7500 but without the four bosses. The material thickness around the two threaded holes for the front cover plate is increased.

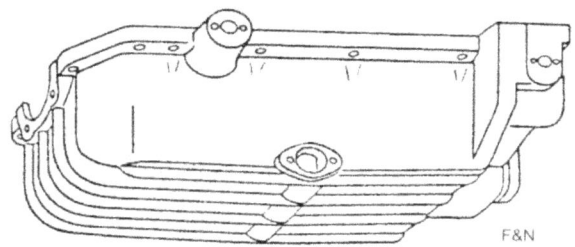

Oil pan 12001 – 14015 As 7501 – 12000 with a reinforcement of the area around the oil return pipe.

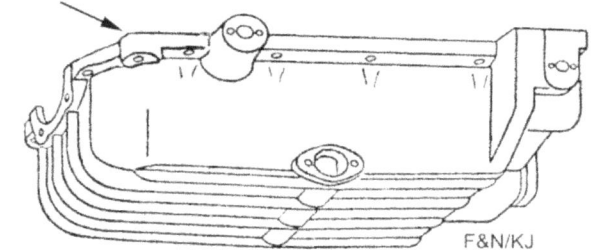

Kick starter
The part of the kick starter that protrudes into the oil pan is the same for all engines. The aluminium kick starter pedal block was unchanged during the whole production period and is identical to the one fitted on the Nimbus-A/B, the »Stove pipe«.
Please note that there is an error in the 1958 spare part catalogue, viz. that the toothed sector (7109) is in reverse!

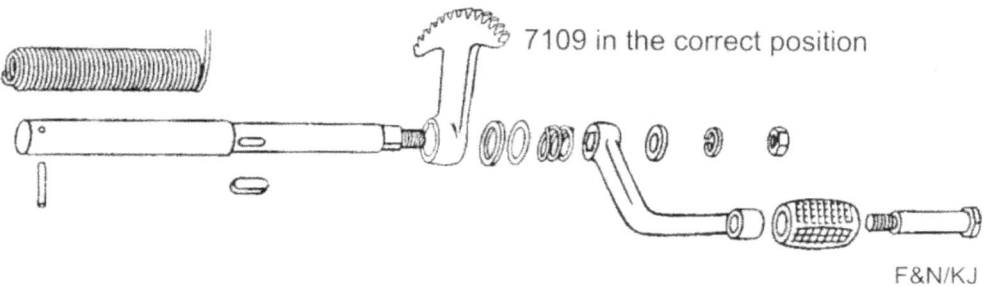

There are some small and unimportant variations in the design of the kick starter arm and the bolt for holding the pedal block.

Bolt for tightening the oil pan and cylinder block together.
- 1301 – 1450 with plain countersunk, flat head.
- 1451 – 7500 with a raised countersunk head.
- 7501 – 14015 with a larger raised countersunk head.

EXHAUST SYSTEM

The exhaust system of Nimbus-C consists of a cast iron exhaust manifold which is tightened onto the cylinder head, a heat shield from nickel plated, later chrome plated, brass, a steel exhaust pipe with or without an integrated fish tail and internal silencer.

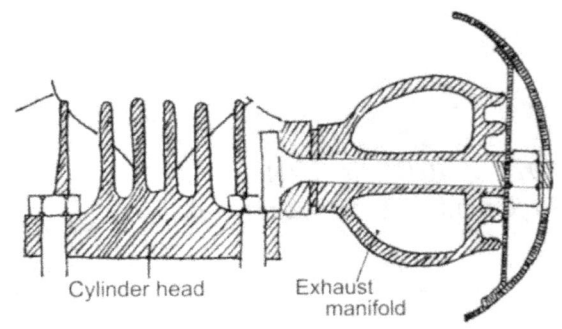

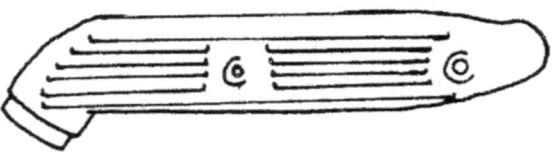

Manifold 1301 – 2550
The top and bottom cooling fins run all the way to the front bolt. By the same token, the adjacent fins run further than the two middle ones, thus forming the shape of a ring around the bolt position.

Manifold 2551 – 14015
As 1301 – 2550, but with all cooling fins having the same length.

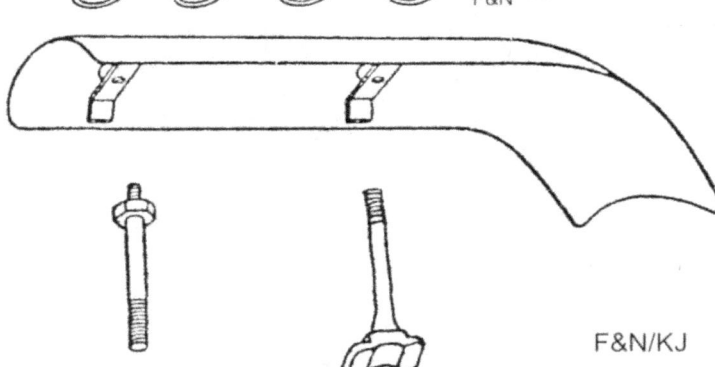

Tightening bolt 1301 – 12880
An M8 bolt with a 14 mm hex head, on top of which is an M6 threaded stud.
The gun metal clamp comes in different versions.

Tightening bolt 12281 – 14015
The clamp and bolts are forged together as one piece. The manifold and the heat shield are kept in place by the same nuts.

Heat shield 1301 – 1550
A nickel plated shield with mildly bent edges fits onto the frame. For a short period of time some of the first machines were running without a heat shield!

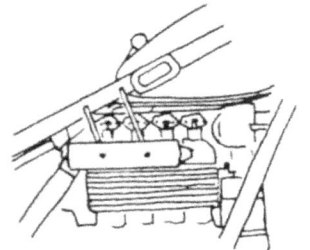

Heat shield 1551 – 12880
Chrome plated brass with two riveted brackets.

Heat shield 2860 – 9000
'Sport' version as after 1551, but with a truncated downwards bent end.

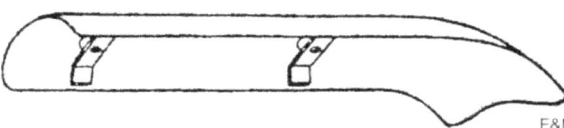

Heat shield 12881 – 14015
As 1551 – 12880, but the holes for the brackets were drilled up to 8,4 mm diameter.

EXHAUST PIPE

There are two different versions of exhaust pipe for Nimbus-C:
- Low level (downwards bent) exhaust pipe
- High level exhaust pipe

Low level exhaust pipe 1301 – 1550
Black enameled short steel pipe with integrated fish tail. In addition, some chrome plated pipes have been found with the same construction.

Low level exhaust pipe 1551 – appr. 4500
As 1301 – 1550 for 'Standard' and Standard Extra versions. Chrome plated for 'Luxus' version.

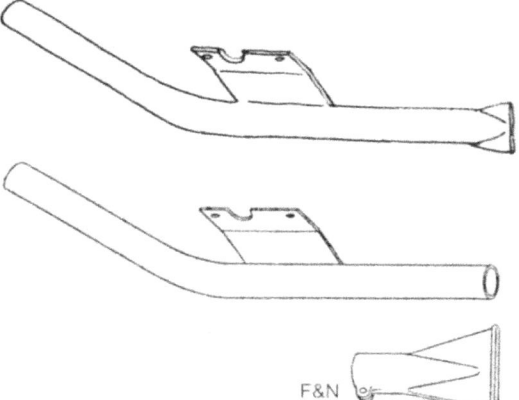

Low level exhaust pipe appr. 4500 – 7500
Black enameled long steel tube with integrated fish tail for 'Standard' version.
The same construction, but chrome plated, for 'Luxus' and 'Special' versions. The tube is a bit longer than for 1301 – 4500 to prevent the rear wheel and rear mudguard from becoming black with soot.

Low level exhaust pipe 7501 – 14015
Black enameled, short steel pipe, without integrated fish tail, for all machines. Separate small fish tail from cast aluminium or chrome plated steel.

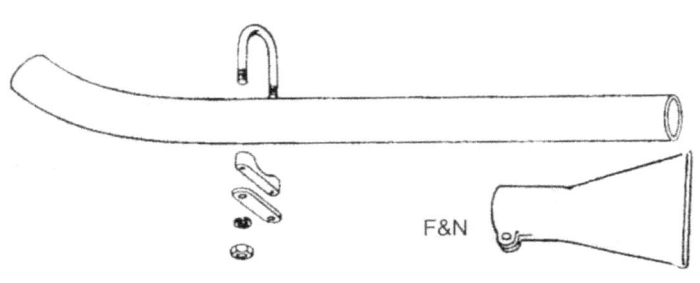

High level exhaust pipe 2860 – 9000
Chrome plated steel pipe, fitted with clamping brackets to fit it onto the frame. Fitted with a really large separate fish tail. The large fish tail was fitted onto the 'Sport' version, but could be fitted to other versions as well.

SILENCER

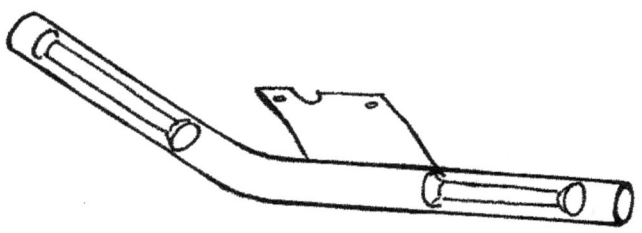

The silencer consists of a perforated internal pipe, centered by two funnel shaped end pieces. The funnels are sealed against the outer pipe with asbestos wool or rope. All silencers are the same, but the exhaust pipes with integral fish tail; the silencer is placed in the front end of the exhaust pipe, whereas in the case of a separate fish tail, the silencer is placed in the rear end, just in front of the fish tail. Instead of asbestos, fireplace sealing rope can be used.

Machines for postal services were provided with two silencers.

GEARBOX

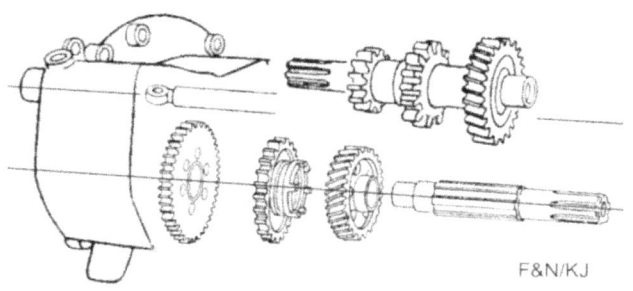

The Nimbus-C gearbox consists of a gearbox housing and –cover, cast in aluminium, fitted with a three-gear gearbox with mainshaft and gear wheels from hardened cast steel.

The 1st and 3rd gears are permanently engaged. The gear wheel for the second gear can slide along the layshaft by the operation of the selector fork and the selector fork rod. , The second gear is engaged in one of the three gears. The gearbox is pressure lubricated by lube oil, directly supplied by the oil pump. The oil returns to the oil pan through a drain hole near the base.

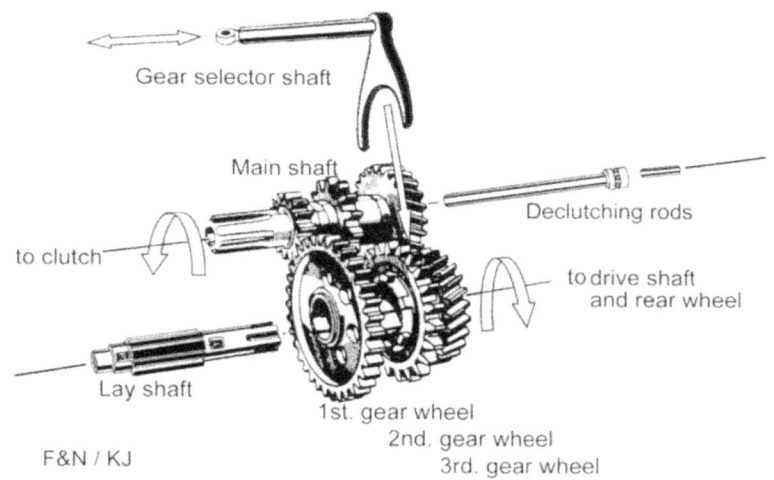

The gearbox comes in two versions, each one with modifications over time.
1. The 'early' gearbox till 7500 with a one point clutch release lever.
2. The 'later' gearbox from 7501 with a two point clutch release lever.

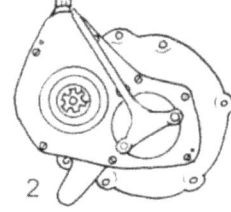

All gearboxes from 1301 – 7500 (the early gearbox) have gear wheels on the mainshaft with 14, 14 and 24 teeth for the 1st, 2nd and 3rd gear respectively.
The corresponding number of teeth of the gear wheels on the layshaft is 34, 29 and 24 teeth respectively.

Gear box 1301 – 1525
The 1st and 3rd gear wheels are provided with rectangular edged gear dog clutch teeth with corresponding slots in the sliding layshaft gear wheel.

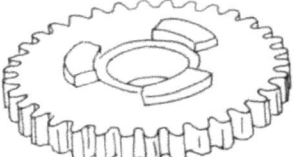 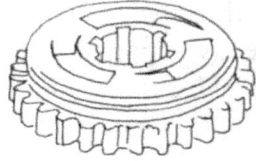 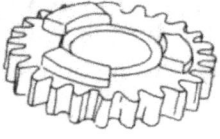

The declutching rod runs through the gearbox block without a gasket. The selector fork rod has four notches.

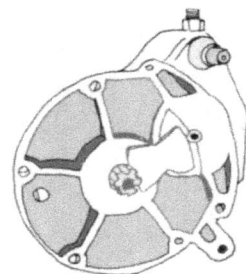

The gearbox housing has no reinforcing near the second fixing hole from the top. At the clutch side, a butterfly shaped casting flash can be seen. The cover is designed to suit the rigid drive shaft. (See Andersen (1996) page 40-41)

Gearbox 1526 – 1550
As 1301 – 1550, but with round gear dog teeth and -corresponding holes in the gear wheels.

Gearbox 1551 – 2050
As 1526 – 1551 but with an increased bore in the mainshaft to allow a cork gasket to be fitted around the declutching rod. The layshaft now has five notches, so there is also a notch between the 2nd and 3rd gear.

Gearbox 2051 – 2560
As 1551 – 2050, but with a hole in the clutch release lever adjustment screw for fitting a split pin.

Gear box 2561 – 7091
As 2051 – 2560, but the collar in the cover for the driveshaft was modified to allow the new cushioned driveshaft to run in the modified collar with a cork gasket.

The pin in the layshaft for the enclosed spring in the former rigid driveshaft became obsolete.
At the same time, the material around the second fixing hole from the top was reinforced.

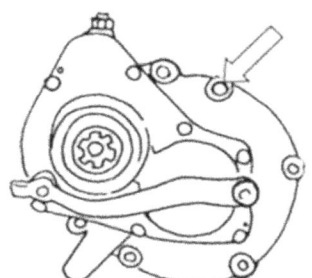

Gearbox 7092 – 7500
The divided declutching rod with separate clutch release thrust balls was made shorter to allow for fitting a ball bearing with five ¼" balls instead.

The gear housing has a cast collar on the bottom to allow the gear wheels for the 3rd gear to run in a local reservoir of oil.

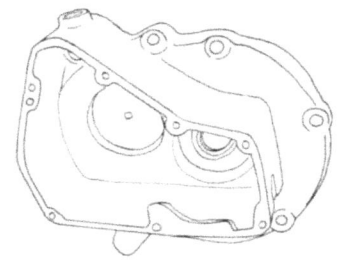

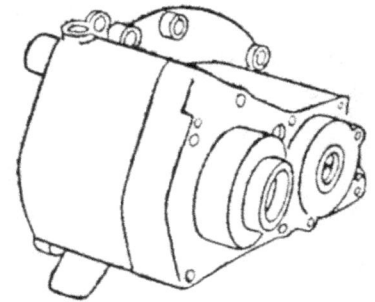

Gear box 7501 – 9000
This gearbox is an intermediate version between the early and the later gearbox; the outside of the housing and the cover are modified and the gear wheels were modified also.
The declutching rod is fitted in two points. The declutching cable is provided with an eye near the threaded plug. The upper part of the threaded plug became obsolete.

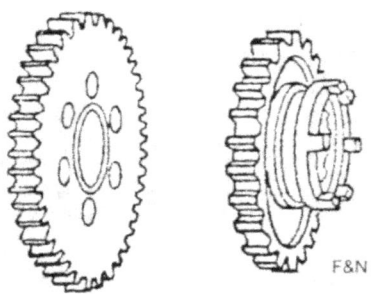

Instead of the three round gear dog-teeth in the 1st and 3rd gear, these gear wheels are now provided with six holes and the 2nd gear wheel with six corresponding teeth at both sides.
The mainshaft still has gear wheels with 14, 19 and 24 teeth for the 1st, 2nd and 3rd gear respectively. The corresponding gear wheels on the layshaft have 34, 32 and 24 teeth.

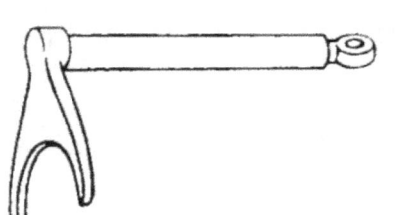

The selector fork rod was modified such that the forks were welded onto the rod. The groove in the selector fork rod was modified to have a threaded hole for the pull rod of the foot-operated gear shift.

At this point in time (around 1950) the factory was obviously highly engaged in improving the gearbox, especially in attempts to reduce the 'howling and singing'. This means that many years later, now and then, constructions of mainshaft and gear wheels were found that deviate from the spare parts list. For example, 3rd gear wheel with angled teeth have been found as early as on the first gear box version. They do not bear the NIMBUS mark and have presumably been produced by a gear wheel manufacturer.

Gear box 9001 – 9686
The later gear box version is launched. The construction is as gear box 7501 – 9000 but the gears of the 3rd gear have angled teeth, intended to reduce 'singing'. For the same reason, the number of teeth of the 3rd gear wheel on the layshaft was increased from 24 to 25. Consequently, the number of teeth of the corresponding gear wheel on the main shaft was changed to 24 teeth. As a result of the angled teeth, an axial force on the layshaft is created. For this reason, the layshaft was reduced in diameter and a small ball bearing was fitted in the front of the gear box against a boss on the lay shaft that was thus formed.

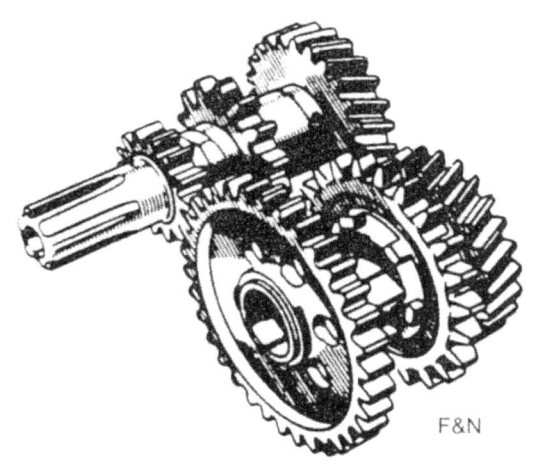

In the gearbox housing, which is capable of accepting the large layshaft bearing, a bushing can be fitted to suit the smaller bearing. It is therefore possible to replace the mentioned parts in gear boxes 7501 – 9000, to obtain the resulting improvements.

Gear box 9687 – 14015
As 9001 – 9686, but the layshaft gear wheel of 1st gear was made a bit narrower and was provided with a collared bushing. This change was presumably made to prevent damage in case of rough shifting from 2nd back to 1st gear.

Pull rod for foot-operated   2860 – 7500
The pull rod has a collar with a hook that is placed in a slot where otherwise the hand-change gear lever was placed.

Pull rod for foot-operated gear shift 7501 – 14015
The pull rod has an eye for a pivot pin (trunion) at both sides.

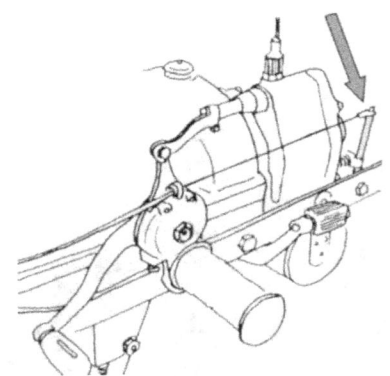

Lever 2860 – 7500
Machines with foot-operated gear shift are fitted with a lever on a pivot mounted onto the frame.

GEAR SHIFT

Nimbus-C can be fitted with a hand- or foot-operated gearshift.

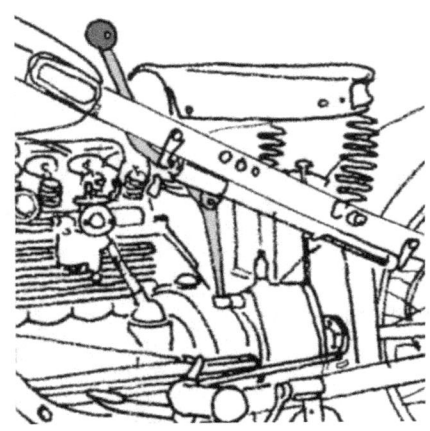

HAND-OPERATED GEARSHIFT

Hand-change gear lever 1301 – 7500
The hand-operated gear shift consists of a hand-change gear lever made from chrome plated special steel with a ball shaped hand grip, consisting of two halves in Bakelite, held together by a bolt at the top end of the lever. There is a lever guiding plate and a pivot pin, as well as a couple of bolts and nuts to attach the construction to the frame.

FOOT-OPERATED GEAR SHIFT

The foot-operated gear shift is a unit that is fitted onto the frame by means of the foot rest at the left hand side.
Machines 2860 – 7500 have the foot-operated gear shift fitted for the versions 'Sport' and 'Special', and as of 1939 (4427) also on the 'Luxus' version. Other machines could be fitted with a foot-operated gear shift for additional premium. By the same token, older machines could be fitted with a foot gear. Housing and lever are normally from malleable cast iron, but in this case the earlier examples are cast in gun metal.

Foot-operated gear pedal 2860 – 8630

The eye on the foot-operated gear pedal is only for the clutch release cable and nothing else. On the pawl disc and –housing, a mark is filed to show the neutral between the 1st and 2nd gear.

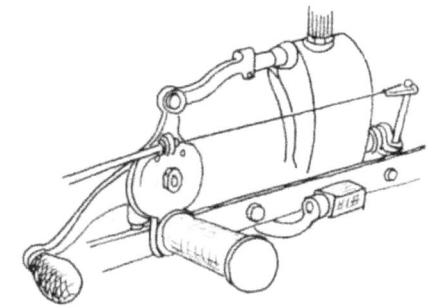

Foot-operated gear pedal 8631 – 14015

The pawl wheel has a stud. There is a bulge on top of the housing which is painted red, to make it easier to see if the gear box is in neutral. The hole for the clutch cable remains, so that it can also be used for older machines.

CLUTCH RELEASE

The clutch of Nimbus-C is released with the left hand and/or foot. Machines with a hand-change gear lever are fitted with both a hand- and foot-operated clutch release, but machines with a foot-operated gear shift have just a hand operated clutch release.

Foot-operated clutch 1301 – 7500

The clutch release pedal is fitted with the left hand foot rest as a pivot. The pedal actuates the clutch by means of an adjustable pull rod, attached to the release lever on the gear box, which in turn actuates the clutch release push rod.
If the machine has a hand-operated clutch release system as well, the clutch cable is attached to an eye in the clutch release pedal.

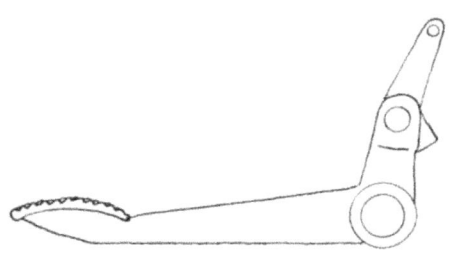

Hand-operated clutch 2860 – 7500
On machines with a foot-operated gear shift, the clutch cable is fitted to an eye in the foot-operated gear housing, and pulls on a lever, which in turn operates the release lever and push rod.

Hand-operated clutch 7501 – 14015
The clutch cable is fitted to an eye on the gear box and pulls on the release lever on the gear box.

DRIVE SHAFT

The power from the gear box to the crownwheel and pinion of the rear wheel takes place through the drive shaft, which comes in three versions.

We cannot really speak of a complex drive shaft as seen on other vehicles, but rather as a simple connecting shaft. This is also true for the two later versions, which have shock absorbers. A more complex drive shaft is able to transmit forces under various, angles under dynamic conditions.

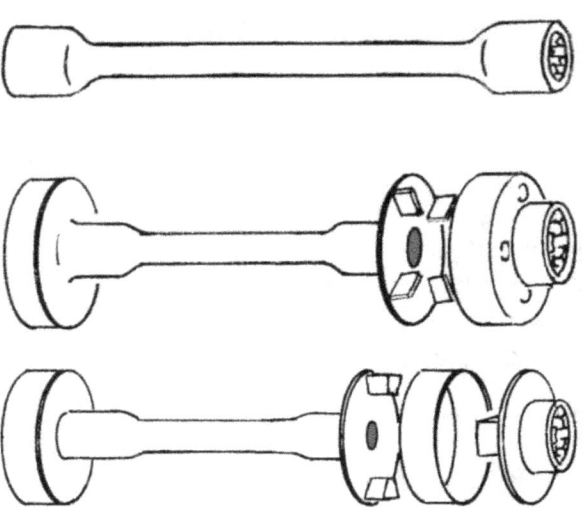

Drive shaft 1301 – 2560
In one piece, often called a 'dog bone'. There is a spring in both ends that presses against a pin at the mainshaft and the pinion wheel respectively.

Drive shaft 2561 – 7091
Consists of an intermediate shaft and two hubs with clearly visible rivets at the ends. Each end of the drive shaft is cushioned internally with eight rubber blocks.

Drive shaft 7092 – 14015
Intermediate shaft and hub are contained in a steel ring at each end. Shock absorbing takes place internally, by means of four rubber blocks at each end.

The rotating drive shaft can, in rainy and dirty weather, contaminate the shoes and trousers of a passenger. Therefore a number of non-original shields have been developed, e.g. an aluminium cylinder that is fitted onto the pinion wheel's cover by means of its four screws.

CARBURETTOR

The carburettor of the Nimbus-C has gone through many changes. At first sight, the differences are small; it is only the change from a small to a large air filter that catches the eye.

The carburettor housing is cast in red brass and is matt nickel pated. All versions are horizontal carburettors with an adjustable needle and an acceleration pump.

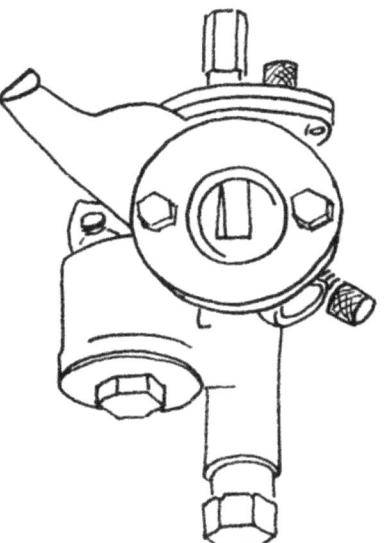

Carburettor '34 – 1 1301 -2487 and 2501 – 2516
The fuel line enters into the top of the carburettor housing. From there, the fuel flows vertically down through a conical valve to the float chamber. To supply fuel at start, a tickler is fitted in the float chamber. The choke consists of a plate that can be turned to restrict and regulate the air supply.
The air suction duct is behind a small screen, which is not really an air filter as such. The carburettor needle is adjusted with a thumbscrew on top. There is another thumbscrew at the back of the carburettor to adjust the amount of air when running idle. The carburettor needle base is pushed up by a spring and the tip is pushed down by the throttle valve; the thinner part is, as usual, near the lower end. The needle is facet tapered.

Carburettor '34-2 2488 – 2500 and 2517 – 3863
As carburettor 1301 – 2487 and 2510 – 2516, but marked "2" on the carburettor housing. The carburettor is provided with a channel that runs from under the throttle valve to the idle running air duct. An adjustment screw was added in the base of the mixing chamber as a lower stop for the throttle valve to adjust the idle running revolutions. The profile of the needle was changed and marked "2". This change was introduced to achieve a more even running when running idle as well as with low rev's.
The change was so effective, that a number of the carburettors from the first series were modified. These carburettors were marked "2" as well on the carburettor housing.

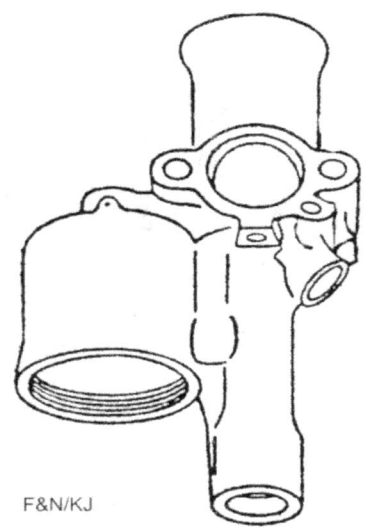

Carburettor '38 3864 – 8500
The factory wrote in a circular: *"... a few machines prior to Nr. 3864 have left the factory with a '38 carburettor."*

Carburettor '38 has an entirely new construction, based however on the same principle. The fuel flows now in from the top. The tickler became superfluous because the conical valve is pretty loose in its nipple, rather heavy and will consequently, by its own weight, follow the downward movement of the float. Contrary to the previous versions, the needle is attached to the base of the throttle valve and the nozzle can be replaced. The acceleration pump has a built-in ball valve. This carburettor is fitted with an adjustable idle speed air screw and an adjustable stop screw for the throttle valve. Approximately from machine 4000 the bore of the main nozzle was increased from 1.3 to 1.5 mm. The reason was, according to the factory, 'to improve running during winter. But the real reason was that nozzle "13" did not supply sufficient fuel! Therefore nozzle "15" was fitted in almost all carburettors.

The fuel mixture can be adjusted by choosing one of the three indentations in the needle. By default, the middle one is used. A three position choke is fitted in the air inlet duct; closed, half open, and open and last but not least, the carburettor is fitted with an air filter with copper mesh.

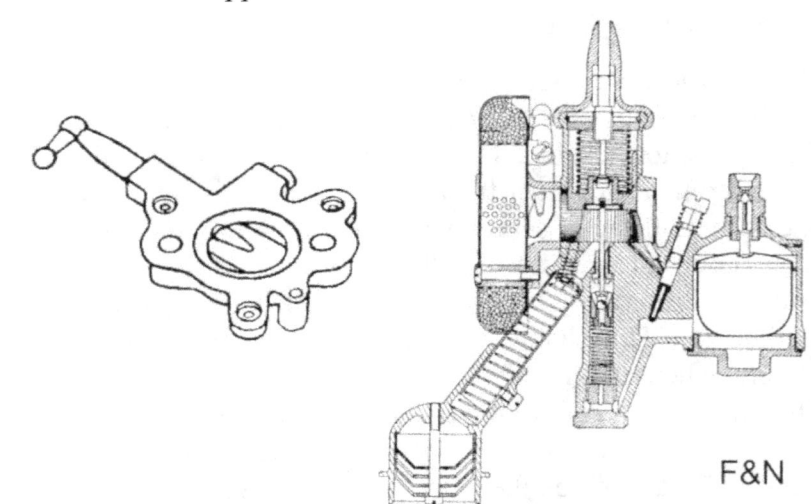

Carburettor '50 8501 – 9200
The dimensions of this carburettor are the same as those of carburettor '38, but the idle running air nozzle and -screw became obsolete to simplify the production. The acceleration pump was simplified by replacing the ball valve by a fiber disc. The main nozzle was not replaceable but pressed into the carburettor's base. The choke's air bore for idle running was deleted.

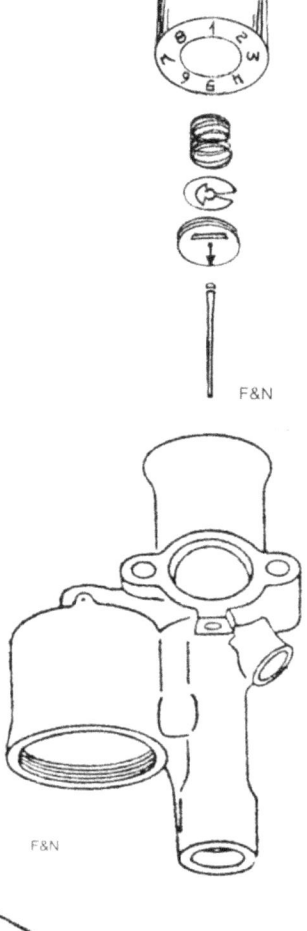

An essential change was made in fitting of the throttle needle. The throttle valve was provided with an internal thread, allowing the needle to be screwed up and down by means of an adjustment disc with a slot. An arrow on the disc points at a number on a ring-shaped scale at the underside of the throttle valve and every single carburettor had its own specific adjustment. This appeared to be a cumbersome construction, driving users, dealers and factory men mad and was therefore quickly abandoned.

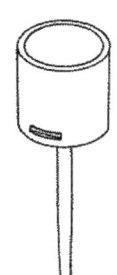

Carburettor '51-1 9201 – 9630
The throttle valve and needle was modified such that the needle was fitted in a small piston in the throttle valve and could be adjusted with an adjustment screw. The throttle needle is the same conical needle with three slots as in carburettor '38 and '50.

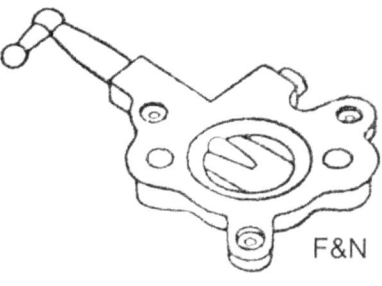

Carburettor '51-2 9631 – 11300
As 9201 – 9630; the adjustable stop screw for the throttle valve moved from the front to the rear (facing the carburettor at the air filter) of the carburettor housing. Consequently, the throttle valve needed to be pushed towards the rear of the carburettor by a spring, to support the draught so that a reliable tightness between the throttle valve and carburettor housing could be achieved. The threaded hole for the former stop screw was closed to prevent air from leaking in.
The carburettor needle was changed from a conical to a faceted needle. Therefore, in order to orient the needle correctly, the throttle valve is provided with a slit with a special M4 guiding screw, fitted in an elevated part of the carburettor housing. The choke of the '50 and '51 carburettor is without an air bore for idle

running. A few '51-2 carburettors have been found having a throttle valve with a slot for an adjustment screw and provided with an adjustment disc as in carburettor '50, but without the ring shaped number scale at the underside of the throttle valve. There is just an adjustment mark.

Carburettor '53 11301 – 14015
With this carburettor the '38 carburettor idle running system was reintroduced. A slanted adjustment screw more or less closes an air bore, which runs from the air filter through the choke to the blending chamber together with a fuel duct that ends in a nozzle in the float chamber. This is in fact a little carburettor in a carburettor!

The removable main nozzle was reintroduced. To be able to distinguish between the nozzle for carburettor '38 (which has an inner diameter of 2.60 mm) and a nozzle for carburettor '53 (with an inner diameter of 2.63 mm), the nozzle for carburettor '53 is marked with a cross. Regrettably this marking stopped as the factory later on started selling needle and nozzle as a set (by the way in a beautiful small wooden box).

Cross matching the nozzles with different dimensions may give problems with achieving the correct mixture.

The adjustable stop screw for the bottom position of the throttle valve is fitted again (as with carburettor '38) fitted at the front of the carburettor, but the position is slanted now, for better accessibility.

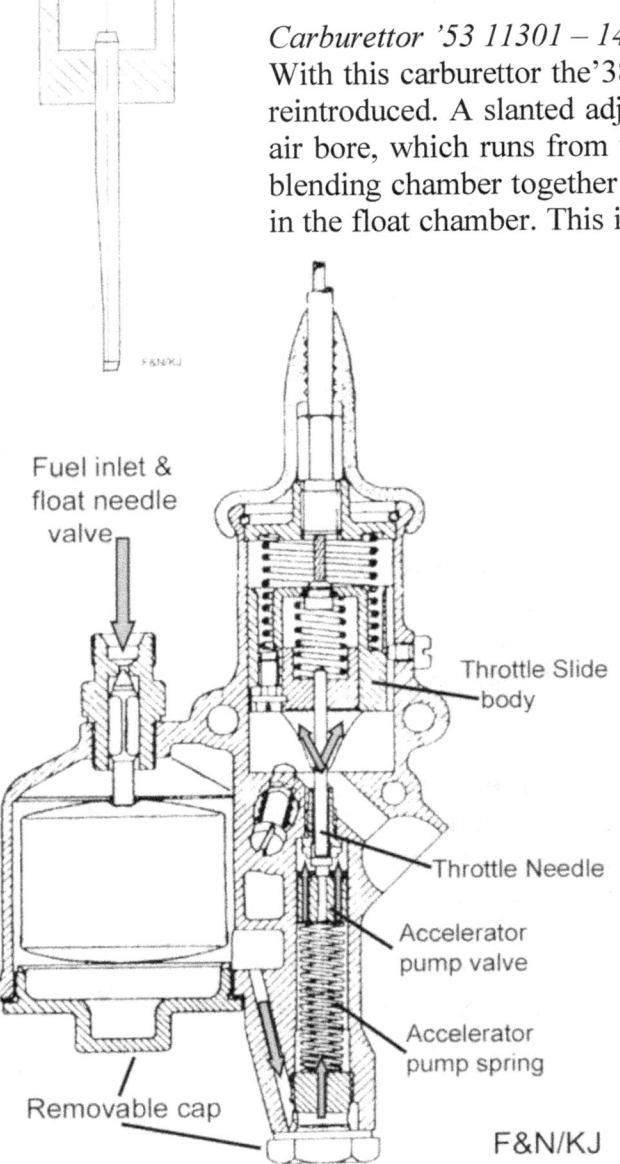

Throttle limiter
To limit Nimbus-C's speed e.g. for vehicles used by the postal services, a steel plate was fitted between the carburettor and inlet manifold. The two bolts for attaching the carburettor had a hole for a throttle plate sealing wire, which made it impossible to remove the throttle plate without being noticed.
To compare, both gaskets are shown.

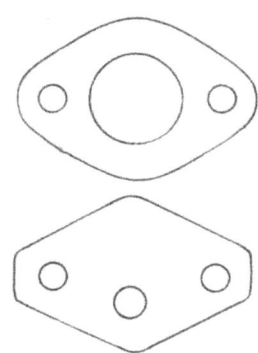

CRANKCASE VENTILATION

Crankcase ventilation is done by the suction force of the carburettor; with Nimbus-C it consists of a matt nickel plated brass tube and an aluminium housing with an oil separator, consisting of two dome-shaped plate dishes, where part of the oil is condensed. The pipe is kept in place between two pipe stubs, a compression spring and a special M4 screw.

The crankcase ventilation comes in two different versions:

Crankcase ventilation housing 1301 – 3863
The pipe stub is cylindrical and the angle is in line with the short ventilation pipe for carburettor '34.

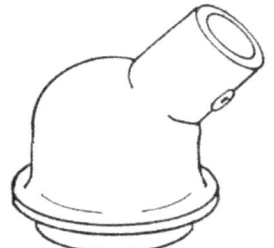

Crankcase ventilation housing 3864 – 14015
The housing is cast in such a way, that the pipe stub is a little conical and has flat sides. The angle is a few degrees steeper compared to the 1301 – 3863 housing and fits the longer ventilation pipe of carburettor '38, '51 and '53.

Crankcase ventilation pipe 1301 – 3863
This ventilation pipe fits onto carburettor '34. The pipe has a crosshatch pattern in the middle and has a 6 mm hole at the underside right under the carburettor to reduce the pressure in the crank case and prevents fuel from running into the crank case housing, in case of a carburettor overflow.

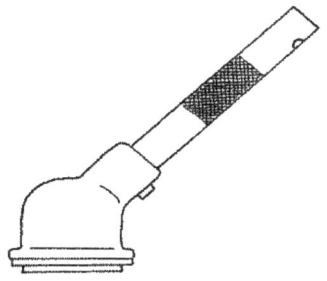

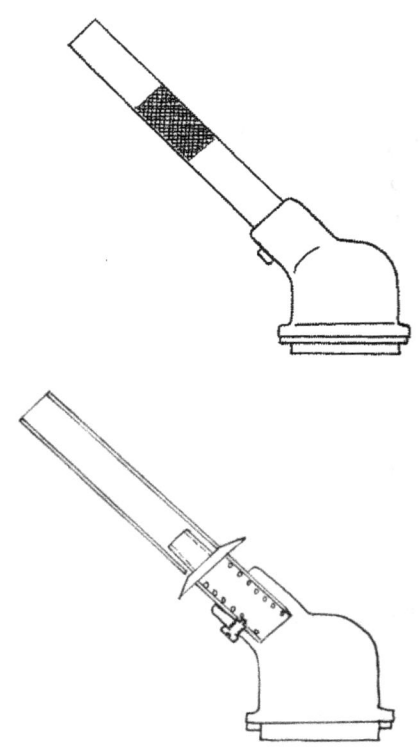

Crankcase ventilation pipe 3864 – 9000
A somewhat longer pipe that fits the '38, '50, '51 and '53 carburettor. Crosshatch pattern in the middle and no pressure-/drain hole.

Crankcase ventilation pipe 9001 – 14015
The pipe is provided with a collar and a drain hole to make sure no petrol will flow into the crankcase in case of a carburettor overflow, but will be drained to the outside instead. The tube with the collar can be found with or without the crosshatch pattern. All pipes with a crosshatch pattern have an indentation at the end of the carburettor. Pipes without the crosshatch pattern and indentation are from the last version.

To achieve oil tightness between the crankcase ventilation housing and the crankcase, a stiffer spring can be fitted into the pipe. But when the inner surface of the housing's pipe stub is worn, this will not be sufficient. The spring pressure will cause the housing to tip from the crank case, leading to oil leakage. Instead of a stronger spring, a horse shoe shaped device can be fitted onto the cylinder block between the two lower cooling fins behind cylinder number four. A -thumbscrew in the arm of the device presses the device towards the cylinder block, resulting in a tight connection between the crank case ventilation housing and the crankcase. This solution (from Jens Bjerregård) was introduced in 1997.

FRONT FORK

The front fork of the Nimbus-C was a telescopic fork from the beginning of development for all versions. The most visible change of the construction of the front fork took place in 1948, when it was changed from a 'low' to a 'high' front fork.

The two main constructions of the front fork are:
- 1310 – 7500: low front fork; the wheel goes up and down within the front mudguard.
- 7501 – 14015: high front fork; the front mudguard follows the movements of the front wheel.

A front fork consists of:
- A lower yoke made from malleable cast iron (1)
- A steel stem (2)
- Two steel fork tubes (3)
    with inner- or outer steel tubes at either end
- And a spring system and two leather or rubber sleeves

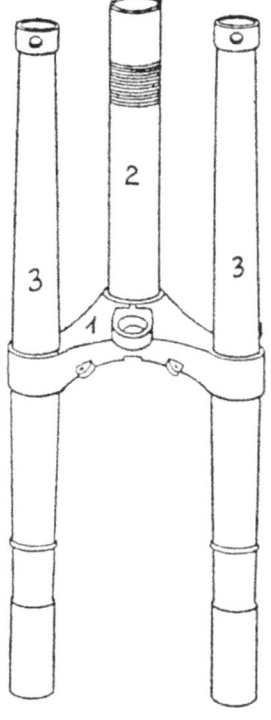

Front fork '34 1301 – 2550
The lower yoke is cast with a head light support and eyes on which to attach the front mudguard. Movements of the two inner fork tubes are guided by a screw in a keyway at the side. At the opposite side of this screw, there is a grease nipple. A long main- and an auxiliary spring are fitted in each tube. The whole assembly is lubricated through the grease nipple. At the lower end, leather sleeves are fitted. From 1400, the two inner fork tubes were modified and the keyways for the screws were made shorter. The bottom blanking plug can have a diameter of 26 mm or, the same as the tube, 25 mm.

Front fork '36 2551 – 3000 and 3135 – 3200
Same construction as 1301 – 2550, but the steering head tube has a key way at the rear to accept a pin on a lock plate, fitted between the threaded bearing cone and the lock nut.

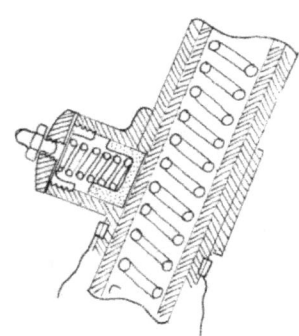

Front fork '37/38 3001 – 3150, 3201 – 4666 and 5000 – 5100
Basically the same constructions as the older one, but both fork tubes are fitted with a T-piece in which a spring loaded bronze plunger is fitted.

These plungers press onto the inner tubes and damp the up and down movement. This construction may sometimes be described as the 'bud front fork'.

Front fork '39 a few before 4667 and 4667 – 4999 and 5101 – 7500
Yoke and stem as before, but with modified inner- and outer fork tubes, a new spring system and rubber sleeves.

Instead of one main spring and one auxiliary spring, there are now three springs with a progressive effect. At the low end inside the outer tube, a brass bush is fitted, which acts like an oil damper valve; in the lower end of the inner tube an oil reservoir is formed, connected externally to the outer tube. Instead of leather sleeves, rubber sleeves were fitted; the lower end has a larger diameter than the upper end. Until 6096 the rubber sleeves were secured with clips, later with wire.

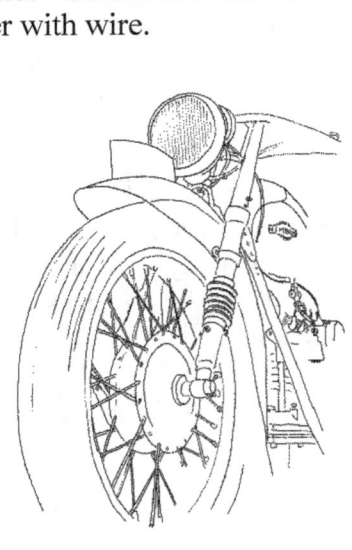

F&N

This front fork can also be used in combination with a front wheel with small brakes. In this case, the spring system contains either the original construction with one main spring and an internal auxiliary spring, or one of the newer constructions with three springs. The left- and right hand side of the inner front fork tubes with the external oil container are identical and have a little oil screw at the end, while on the 'real' '39 front fork, they are positioned at the front.

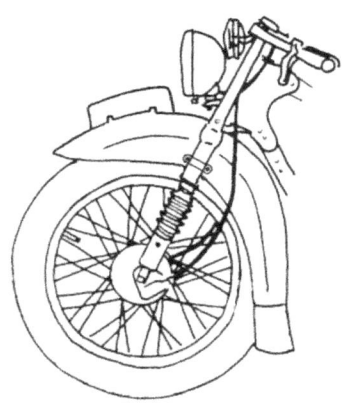

Some front forks from that time have a construction that looks like the new '48 version. They are fitted with a front wheel with small brakes and a low front fork, but the front mudguard follows the up and down movements of the wheel. We do not know which motorcycles were fitted with this construction.

Another combination of front fork and –wheel is a '35, '36 or '37/38 front fork with a front wheel having a 180 mm brake fitted. Aluminium anchor plates have also been found in this combination.

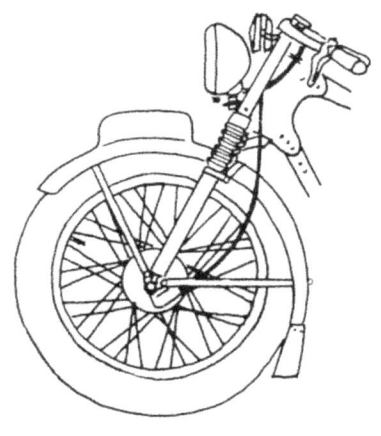

Last but not least, photographs have been found showing a front fork with a support for the horn at the place of the headlight in one piece. In this case, the headlight is fitted onto two brackets, at either side, as with the front fork after 7501.

Front fork '48 7501 – 8814
With this version of the front fork, the telescopic construction was changed. The front wheel and front mudguard move up and down together. The yoke, the steering head tube and the angle between them and the front fork tubes were modified, and the caster was changed from 60 to 65 mm. (The caster is the distance between the imaginary point on the ground where the centre line through the yoke and steering shaft hits the ground and the point where the wheel touches the ground. The larger the caster, the better the track stability, but the harder it is to change direction).

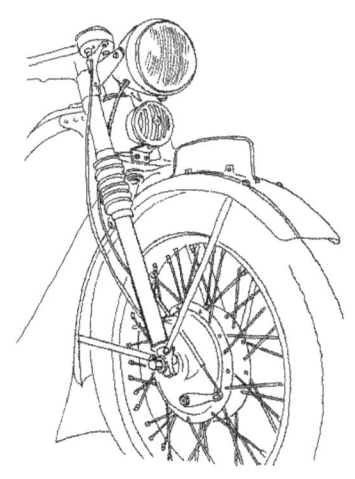

143

From this point it is possible to change the fork tubes individually, as they are now fitted in the yoke with special bolts. The rubber sleeves are changed as well; they have the same diameter at both ends and can be secured with iron wire. The spring system is the same as with the '39 front fork: three progressive compression springs. In addition, an extra spacing tube was fitted, because the '48 front fork legs are longer than those of the '39 front fork.

Front fork '48-2 8815 – 9000
The same construction as 7500 – 8814, but the nuts of the bolts in the yoke changed from spanner size 17 to 14 mm, presumably to avoid overstressing the fork tubes.

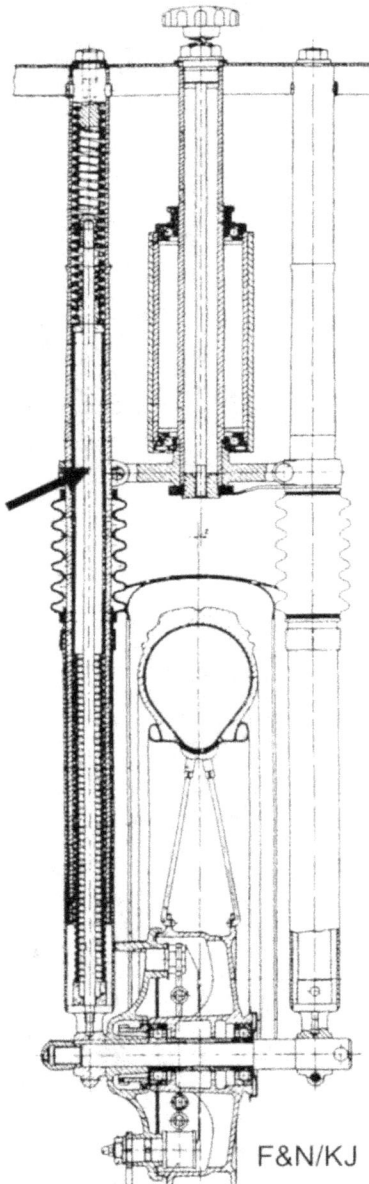

Front fork '50 9001 – 14015
Seen from the outside, it looks the same as the '48 front fork, but the spring system is modified. The middle spring is replaced by a steel distance tube with the same diameter as the springs. The distance tube also serves as a safety device, because in the event of a collision, the front fork legs near the yoke would otherwise have the highest chance to break off (see arrow on adjacent drawing). The lower spring comes in two versions: the older version for sidecar- and the less robust version for solo machines. The upper spring, the distance tube and the lower spring define the length of the front fork. Therefore, the lower distance tube became obsolete. Finally, a bronze plunger valve was fitted in the bottom of the outer tubes, to prevent 'slamming' because the oil pressure will steeply increase just before the front fork tubes hit the bottom.

F&N/KJ

HANDLEBARS

The handlebars of Nimbus-C come in two steel plate versions, one for each front fork construction: handlebars with instruments for the low front forks and those without instruments for the high front fork.

Handlebars 1301 – 1550
Flat handlebars from 2.0 mm plate steel with riveted twist grip rings. The rivets are visible on the upper side of the handlebars.

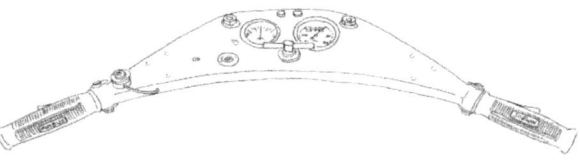

Handlebars 1550 – 4600
As 1301 – 1550, but with countersunk rivets which cannot be seen on the enameled surface.

Handlebars 4601 – 7246
The handlebars are no longer fully flat, but both twist grips and the extreme ends are bent slightly upward, about 5°. The ring shaped twist grip retainers are now welded instead of riveted onto the bottom side of the handlebars. The plate thickness was increased from 2.0 to 2.5 mm.

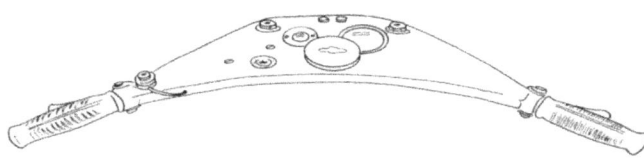

Handlebars 7247 – 7500
As 4601 – 7246, but the hole for the speedometer at the right hand side is closed. The ends of the handlebars are bent upwards to approximately 10°.

Handlebars 7501 – 14015
There are now holes in the handlebars for fitting the high front fork. At the right hand side, the twist grip is fitted into a ring with 28 pieces of 1/8" steel balls. There is a hole for the charge indicator lamp, belonging to the new combination switch.

The two versions of handlebars are fitted with the following parts and devices:

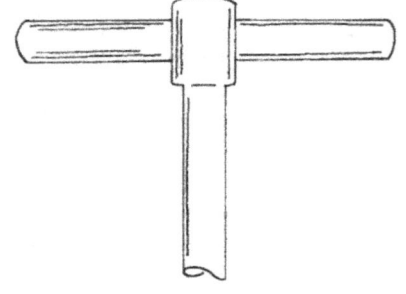

Steering damper assembly

Adjustment
1301 – 2600: With T-bar, chrome plated.
2601 – 7500: With Bakelite grip.
7501 – 14015: As 2601 – 7500, but with a 50 mm longer threaded rod.

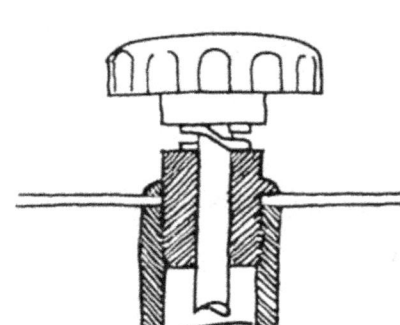

Double spring discs (Thackeray washer)
1301 – 6400: Not present at that time, but can be fitted.
6401 – 14015: As shown.

Threaded plug (in place of damper)
1301 – 2600: Matt nickel plated, 19 mm across flats.
2601 – 14015: Cadmium plated, 27 mm across flats.

Bottom clamping disc:
1301 – 7500: Matt nickel plated disc.
7501 – 14015: Cadmium plated disc with four indentations, corresponding with the notches in the head tube.

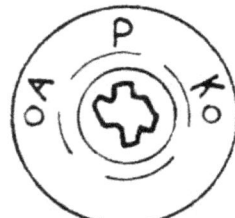

Ignition switch surround
1301 – 7500: Chrome plated brass with the characters 'A', 'P' and 'K';
'A' for off (Afbrudt), 'P' for parking (Parkering) and 'K' for run (Kørsel).

7501 – 14015: Chrome plated without characters.

Lighting twist grip
1301 – 7500: Steel pipe, length 136 mm inside the handlebars and 125 mm twist grip, diameter 24 mm.
7500- 14015: Steel pipe, length 123 mm inside the handlebars and 125 mm twist grip, diameter 24 mm.

Throttle twist grip
1301 – 7500: Steel pipe, length 136 mm inside the handlebars and 125 mm twist grip, diameter 24 mm.
7501 – 14015: Steel pipe, length 196 mm inside the handlebars and 125 mm twist grip, diameter 24 mm. Fitted into two ball bearings with 28 pieces 1/8" balls.

Rubber handgrips
1301 – 4600: Conical chamfer both ends with an area where NIMBUS is moulded into the rubber.

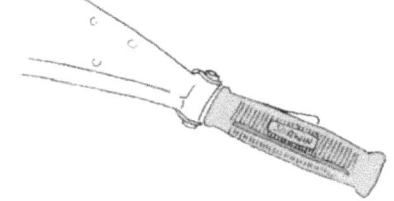

4601 – 9000: Conical chamfer both ends with a firm bulb at one end which fits inside the handlebar tube.

9001 – 14015: Cylindrical with a fine cross pattern with an area where "Nimbus" is moulded into the rubber.

There are also handgrips with NIMBUS moulded in the rubber using smaller characters, and rubber handgrips without the NIMBUS field at all.

For horn push button, charging indicator light and combination switch, see chapter »Electrical system«.

Levers
Clutch- and brake levers have the same shape for all machines, but the material differs.
1301 – 155: Chrome plated steel.
1551 – 14015: Mostly stainless steel, sometimes chrome plated brass. Chrome plated steel levers were fitted during periods that stainless steel and brass plate was hard to get.

Screws for levers
1301 – appr. 6400: M6 x 0.75 x 5.5 mm long screws with a 10 mm hex head.

Appr. 6400 – 14015: Special M6 x 0.75 x 5.5 mm long slotted round-head screw with a 0.8 mm relief on the diameter of the head to provide a boss for the lever.

Speedometer
See chapter »Speedometer«.

WHEELS

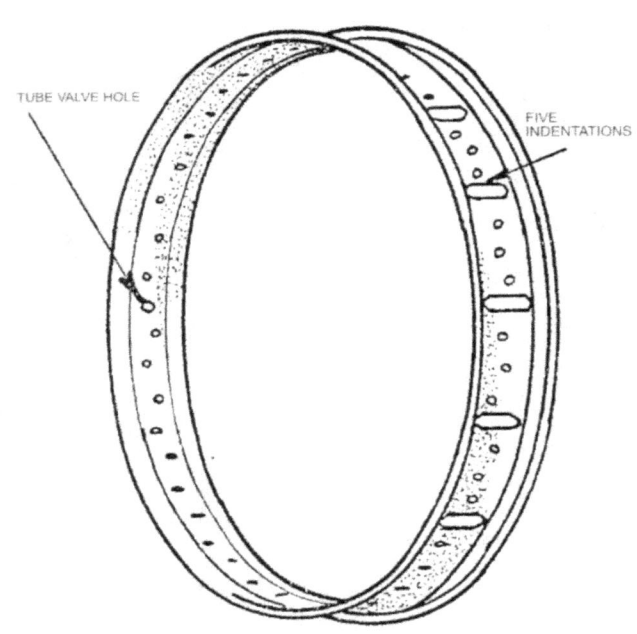

Basically, all Nimbus-C wheels are the same. Both, front- and rear wheel have steel rims, formerly referred to as WM3" x 19", which is now 2.15" x 19". The rims can be enameled or chrome plated and could later be purchased in stainless steel. There are 40 holes for the spokes and one for the tube valve. The 19"x 3" rim called for a tyre size of 3.50"x 19" or possibly 3.25 x 19".

During appr. 1938 – 45 so called safety rims were used; rims with a row of indentations in the base of the rim over half the circumference, to prevent the tyre from slipping.

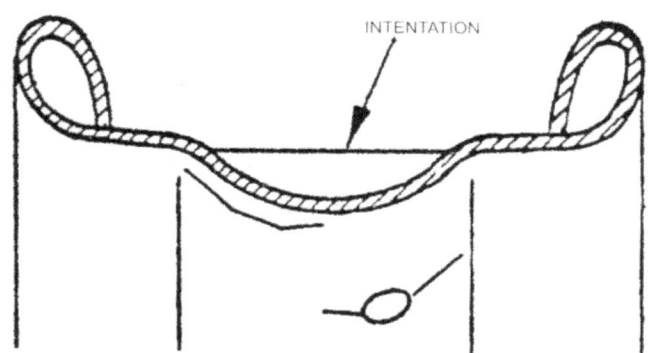

All Nimbus brakes are drum brakes. The brake drums with a diameter of 150 mm are riveted onto the front- and rear hub. Brake drums with a diameter of 180 mm at the rear wheel are bolted to the hub and those for the front wheel are cast as a unit together with the front wheel hub (full width hub brake).

FRONT WHEEL

A front wheel consists of:
- Rim, rim tape, tyre and inner tube
- Spokes and nipples
- Hub with brake drum
- Anchor plate, brake shoes, springs, eccentric cam, cable and lever
- Speedometer drive
- Front wheel and bearings

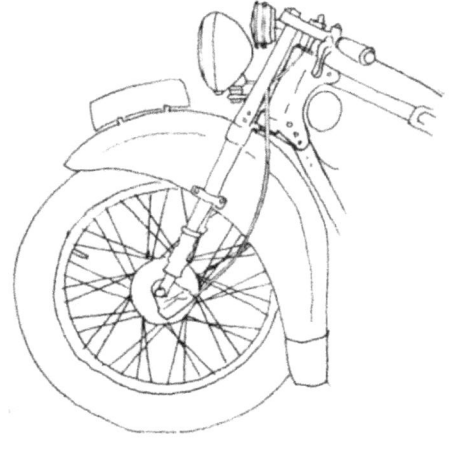

Front wheel 1301 – 4601 (some until 4666) and 5000 – 5100
150 mm enameled brake drum, riveted onto the left-hand side of the hub, both made from malleable iron.

Front wheel 4602 – 4666
Some of these machines have a 150mm brake drum fitted at the left-hand side and others ('Sport' and 'Special' in 1939) with 180 mm brake drums over the whole width of the hub.

Front wheel (Some from 4602) 4667 – 4999 and 5101 – 8500
180 mm wide brake drum from enameled malleable iron over the whole width of the hub (full width hub brake)

Front wheel 8501 – 14015
Same front wheel as before 8501, but the setting up of the spokes has changed; every spoke crosses three other spokes now. The new front wheel spokes have a kink near the rim,

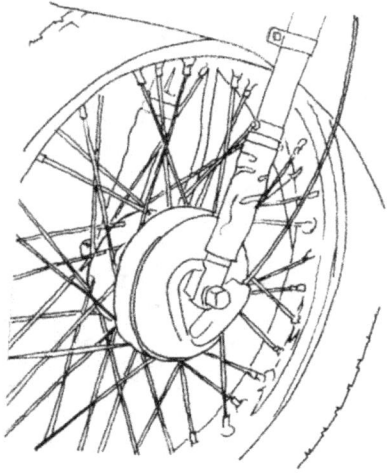

Front wheel brake 1301 – 1525
The brake anchor plates are made from malleable iron and are flat. The brake is operated by an eccentric cam and effects the rollers of the brake shoes. There is a coiled torsion spring against the eccentric cam and a pullback spring for the brake shoes.

Front brake 1526 – 2550
As 1301 – 1526, but the leg end for the fork tubes was made 15 mm longer, and the diameter of the centre bushing was increased from 26 to 27.5 mm. The hole in the eccentric cam was changed accordingly. (See Andersen (1996) page 95)

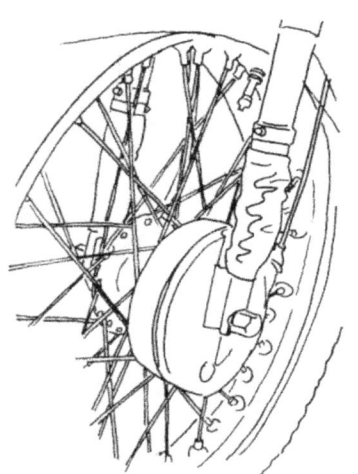

Front brake 2551 – 4601 (Some until 4666) and 5005 – 5100
The brake anchor plate was modified to have a right angled rim, so water and dirt could no longer enter into the brakes. The eccentric cam and the coiled torsion spring were both made more robust. The contact surface between the brake anchor plate and the fork tubes was made longer and the cable was fitted with a rubber hat.

Front brake (Some from 4602) 4667 – 4999 and 5101 – 7500

The enameled brake anchor plate, made from malleable iron, is fitted at the right hand side and forms the enclosure for the speedometer drive. There is a fixed pivot for the brake shoes. This model of brake shoes also comes in aluminium. At the left hand side, the bearing is protected against water and dirt with a felt ring, which was later replaced by a protective cap.

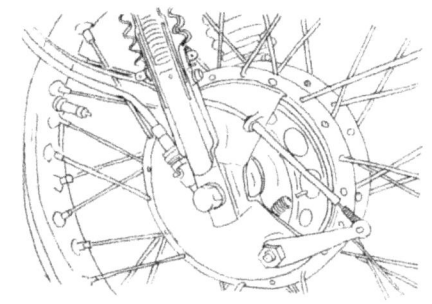

Front brake 7501 – 13850

The brake anchor plate was changed to fit the '48 front fork. At the same time, the brake shoes were changed so they became self-centering. This was achieved by no longer using the eye on the brake shoe and also the pivot rod in the brake anchor plate; a contact stud in a threaded hole in the brake anchor plate was fitted instead.

Front brake 13851 – 14015

The contact surface between the outer tube of the front fork and the brake anchor plate was considerably reinforced. For the rest, the construction remained the same as after 7501.

Speedometer drive 1301 -2050

A gearwheel (crown wheel) with 36 teeth is fitted on the front hub and a pinion with 12 teeth is fitted in the bushing of the enameled enclosure, made from malleable iron. The enclosure is fitted at the right hand side of the front wheel. The pinion drives the speedometer cable with a reduction of 1:3 (W=1.5).

The bushing is screwed in the enclosure with an M 18 x 1.5 mm thread; however, the thread of the union nut of the speedometer cable is M 17 x 1.5 mm.

Therefore there are bushings (as a spare part) with an M 18 x 1.5mm thread at both ends, to fit a common speedometer.

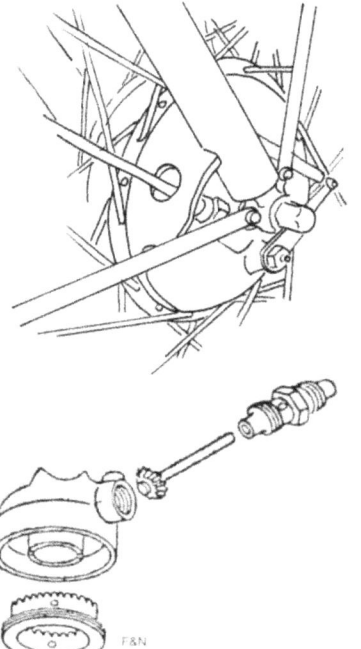

Speedometer drive 2051 – 2900
As 1301 – 2050, but the enclosure is fitted with a grease nipple.

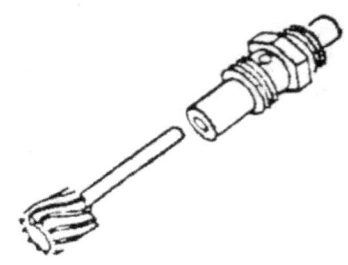

Speedometer drive 2901 – 4601 (Some until 4666) and 5000 -5100
The speedometer drive changed from a reduction ratio of 1:3 (W = 1.5) to 1:2 (W = 1.0). This was achieved by changing the gearwheels; the crown wheel was changed into a worm wheel and the pinion into a worm. That resulted in changed dimensions of the enameled enclosure, made from malleable iron, fitted at the right hand side. The bushing for the worm has an M 18 x 1.5 mm thread at ends, one end to be screwed into the enclosure and the other end to connect onto a union nut.

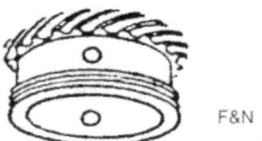
F&N

There are bushings (as spare part) with an M 17 x 1.5 mm thread to be screwed into the enclosure, and an M18 x 1.5 thread for the connection onto the union nut, like the original bushes, but these are now marked with a triangle for identification.

Speedometer drive (Some from 4602) 4667 – 4999 and 5101 - 7246
Same worm, worm wheel and bushing as after 2901, but the enameled brake anchor plate from malleable iron, fitted at the right hand side is shaped such, that it includes the enclosure.

Speedometer drive 7247 – 9500
The same worm wheel as after 2901, but worm and bushing are designed to accept a "Smiths" speedometer with a shorter worm and bushing and a WFG 26 tpi thread for the union nut of the speedometer cable.

Speedometer drive 9501 – 14015
Same worm, worm wheel and bushing as after 2901. The brake anchor plate at the right hand side forms the enclosure for the speedometer drive.

REAR WHEEL

A Nimbus-C rear wheel consists of:
- Rim, rim ribbon, tyre and inner tube
- Spokes and nipples
- Hub and brake drum
- Brake anchor plate, brake shoes, springs, eccentric cam and lever
- Crown wheel and pinion with bearings, fitted into the gearwheel housing with cover
- Rear axle with bearings

Rear wheel 1301 – 2560
Rear wheel 1301 – 1550 is fitted onto the frame with regular nuts on the rear axle shaft. The crown wheel housing has a threaded hole for grease lubrication and one threaded hole in the centre for attachment to the frame. Thin 50 mm hub; ball bearings on both sides and a thrust bearing on the side of the crown wheel. From 1551 all rear wheel axle shaft nuts are closed (blind).

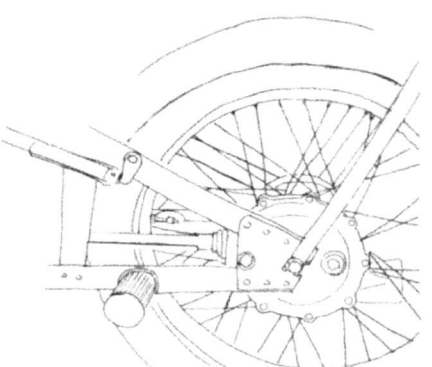

Rear wheel 2561 – 3000 and 3151 – 3196
The threaded stud in the drive housing is replaced by a grease nipple and there are two threaded holes in the housing for attachment to the frame. The profile of the housing is changed and is more rounded. The housing for the pinion is modified, to fit the sprung drive shaft.

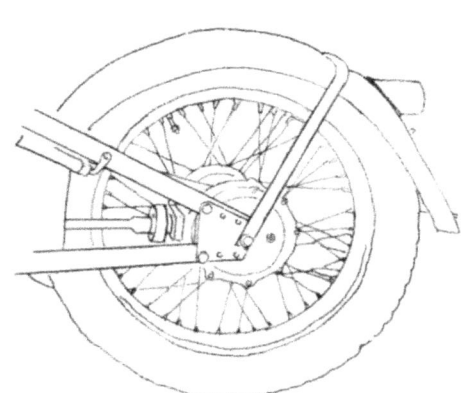

Rear wheel 3001 – 3150 and 3197 – 7500
The hub originally had a diameter of 50 mm, but was changed in the course of time to 75 mm. The wheel runs in two conical ball bearings.

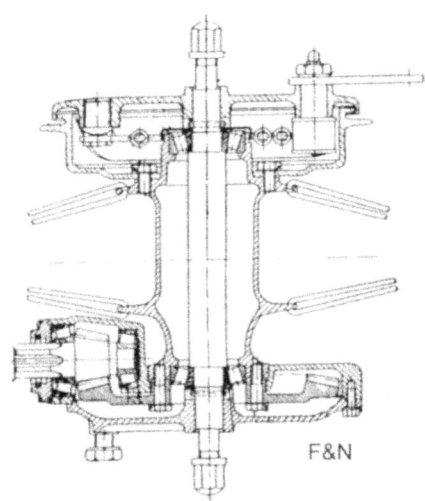

223 Rear wheel 7501 – 14015
The gearwheel housing has its lubrication point on top, closed with one M6 bolt. (The grease nipple became obsolete). During a short period of time, from 1948 – 49 the gearwheel housing as well as the housing was made from polished aluminium. Later, cast iron was used again.

Rear brake 1301 – 2560
150 mm brake drum from enameled malleable iron, riveted onto the right hand side of the hub. The brake anchor plate and the eccentric cam were reinforced starting with no. 1526, in line with the front brake, (See Andersen (1996), page 96f)

Rear brake 2561 – 3000 and 3151 – 3196
Brake drum and hub as 1301 - 2560

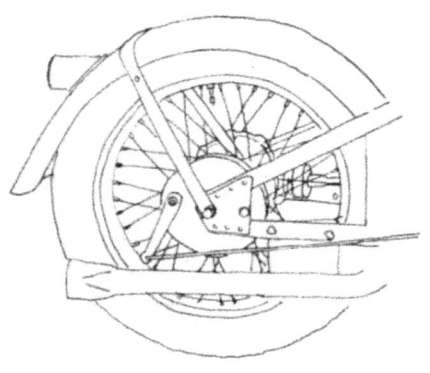

Rear brake 3001 – 3150 and 3197 – 7500
Brake drums 180 mm, bolted onto the rear hub. Brake drum is pressed from steel and enameled.

Rear brake 7501 – 14015
The enameled brake drum is 180 mm, made from malleable iron, at first with a number of reinforcement ribs. From about 1950 there is only one robust cooling- and reinforcement rib. The brake anchor plate is prom polished aluminium and the brakes are self-centering, just like the front brakes.

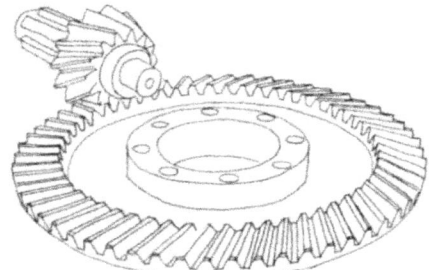

Pinion 1301 – 2560
Pinion made from hardened malleable steel, provided with a drive pin to transmit the power for the rigid drive shaft (See chapter 'Drive shaft'). The pinion is fitted in the final drive housing with a small ball bearing journal, and at the end of the drive shaft, with and a large conical roller bearing.

Pinion 2561 – 14015
Without driving pin and supported in two conical roller bearings.

Crown wheel and pinion
The crown wheel is from hardened forged steel and can have 56 or 59 teeth.
From 1954: 57 or 59 teeth. Apart from the number of teeth, the differences between a crown wheel with 56 or 59 teeth can be seen from the way the 'rear end' has been processed. The crown wheel for sidecar riding with 59 teeth is milled, whereas one with 56 teeth is un-milled.

The solo-crown wheel (from 1954) with 57 teeth and the crown wheels for sidecar use with 59 teeth are processed the same way (and have the same dimensions).

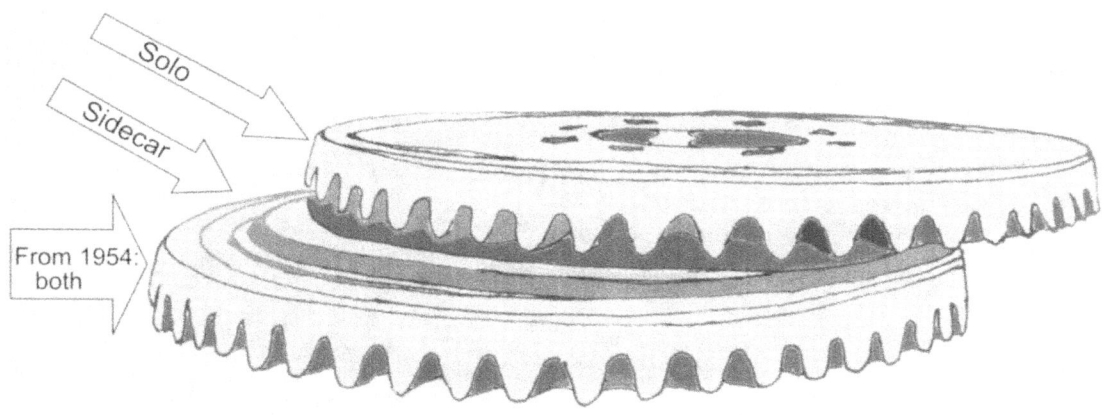

In the 1980's, crown wheels and pinions have been developed with helical teeth for both solo- and sidecar use.

TRANSMISSION

With Nimbus-C there are three different gearing ratios for the gearbox (See chapter 'Gearbox') and three sets of crown- and pinion wheels. This results in nine different ratios. The table shows the six relevant ratios.

NIMBUS-C TRANSMISSION

Rear wheel: Gear box: Number:	Sidecar-gearing: (12 : 59) 1 : 4,9	Solo-gearing: 1301-7500: (14 : 56) 1 : 4	Solo-gearing: 12200-14015: (14 : 57) 1 : 4,07
1301-7500:			
1st gear **1 : 2,43**	1 : 11,9	1 : 9,72	
2nd gear **1 : 1,53**	1 : 7,5	1 : 6,12	
3rd gear: **1 : 1**	1 : 4,9	1 : 4	
7501-9000			
1st gear: **1 : 2,43**	1 : 11,9	1 : 9,72	
2nd gear: **1 : 1,57**	1 : 7,69	1 : 6,28	
3rd gear: **1 : 1**	1 : 4,9	1 : 4	
9001-14015			
1st gear: **1 : 2,43**	1 : 11,9		1 : 9,89
2nd gear: **1 : 1,57**	1 : 7,5		1 : 6,39
3rd gear: **1 : 1,04**	1 : 5,12		1 : 4,23

12:59, 14:56 and 14:57 mean that the pinion has 12 or 14 teeth and the crown wheel has 59, 56 or 57 teeth.

Gearing ratio 1301 – 1550
All machines have a gearing ratio of 14:56.

Gearing ratio 1551 – 7500
Normally for 'Standard' machines 12:59, called 'low gearing' or 'sidecar gearing'. For 'Luxus'-, the later 'Sport'- and 'Special' versions, the gearing ratio is 14:56, called 'high gearing' or 'solo-gearing'. All machines could upon request be delivered with high or low gearing.

Gearing ratio 7500 – appr. 12200
All machines have a gearing ratio of 12:59.

Gearing ratio appr. 12200 – 14015
Normally for 'Standard' version 12:59, called 'low geared' or 'sidecar geared'. For 'Luxus' version, gearing ratio 14:57, called 'high geared' or 'solo geared'. All machines could upon request be delivered with high- or low gearing.

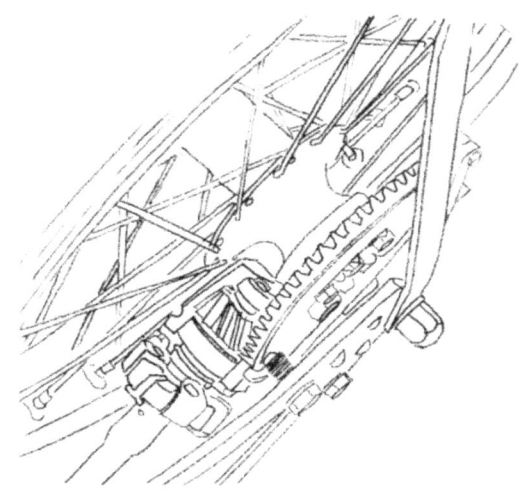

Please note that in 1934 all machines were fitted with crown- and pinion wheels with a gearing ratio of 14:56 and that from 1948 till 1954 only machines with a ratio of 12:59 were built.

BRAKES

All Nimbus-C motorcycles have a hand-operated front brake and a foot-operated rear brake.

FRONT BRAKE

The front brake consists of a chrome plated steel- or brass brake lever, at the right hand side of the handle bars, a brake cable, a brake drum, which is either riveted onto, or built into the front wheel hub and a brake anchor plate with brake shoes with riveted or vulcanized linings (see chapter Front wheel).

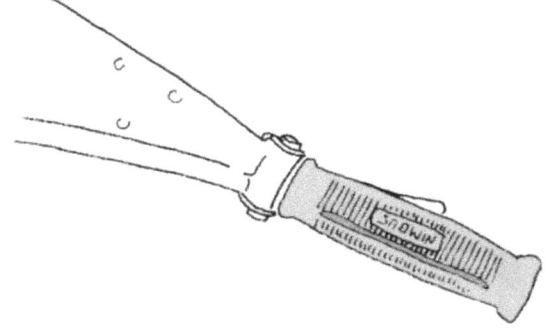

REAR BRAKE

The rear brake consists of an enameled or, later, a cadmium plated brake pedal from malleable steel, fitted onto the frame with the right hand foot rest as a pivot point, with a 2mm thick washer between pedal and pull rod. The pull rod operates a brake lever and subsequently brake shoes with riveted linings, fitted on a brake anchor plate. The brake drum is fitted onto the rear hub by either rivets, or screws (See chapter Rear wheel).

Brake pedal 1301 – 3000

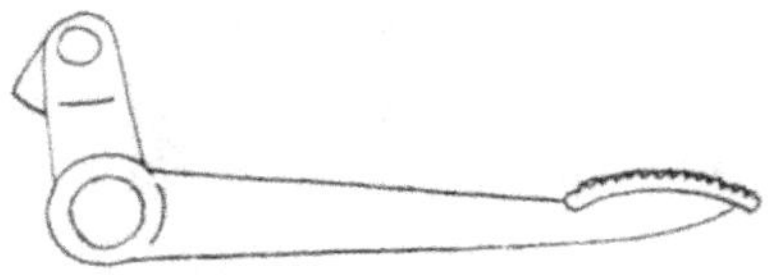

Brake pedal cast in special steel, about 150 mm long. It has a stop at the rear against the frame rail. Because at that time, sidecars for NIMBUS-C were without brakes, there was no hole in the brake pedal for the brake pull rod for the sidecar.

Brake pedal 3000 – 7500

A longer (about 170 mm) and more robust brake pedal, cast in special steel, with an additional hole for the sidecar brake rod.

Brake pedal 7501 – 14015

As 3001 – 7500, but with a reinforced angle section and a more robust construction.

Brake rod 1301 – 3000
700 mm long, diameter 8 mm, enameled or, later, cadmium plated steel.

Brake rod 3001 – 14015
780 mm long.

Brake drum, -shoes and –anchor plate
See Chapter Rear wheel

Brake eccentric cam, etc.

Since Nimbus-C was fitted with 180 mm rear brake drums from 1937 on, the brake construction was fitted with an eccentric cam, named "Brake eccentric cam for the rear wheel". (Note: At this time the front wheel was fitted with a 150 mm brake).

The distinguishing feature of the earliest eccentric cam was that the threaded section was long enough to accept the lever retaining disc, the brake lever, and a thick nut with spring washer. Furthermore, there is just one hole in the eccentric cam spindle for greasing and there is no lubrication groove.

The brake lever does not have the mark "S", and can be fitted either way around on the shaft (flipped). The stamped-out square hole in the lever is symmetrical and hence has the same operating angle in both orientations. The retaining disc between the brake lever retaining nut and the brake lever is the same beveled disc (4102) that is fitted under the clutch- and brake pedals. The nut (7261) is the same as the one used for the foot rests.

In 1939 the front wheel of the 'Sport' and 'Special' versions were also fitted with a 180 mm brake, but its eccentric cam had a shorter threaded end, and consequently a thinner nut (8683).

Between the brake lever retaining nut and the brake lever, the steel disc was changed to a felt gasket, impregnated with talcum instead. The brake lever is shorter than that of the rear brake and has a punched hole such, that different operating angles are possible, and depending on which way the lever is fitted.

If 180 mm brake linings are worn but still usable, and brake arm travel is excessive, a first step would be to remove the brake arm (8682 front, or 8380 rear) and turn it over so that the "S" ("S"= Slidt, worn) stamping is facing outward.

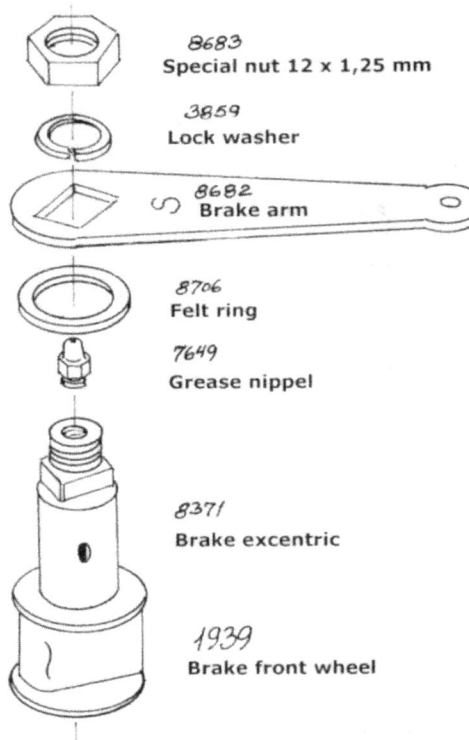

So, if an "S" on the brake lever is facing outward, this indicates that brake linings are partly worn. Note that the eccentric cam has still one lube hole without a grease groove.

Early in 1939 information from the factory was received that rear- and sidecar brakes were also to be marked with an "S" and could be flipped over.

At the same time, a different construction of the rear brake lever with an octagonal hole was used with a corresponding spindle. Although this construction is not recorded in the F&N archives, the complex shape indicates that is was most likely produced by the factory.

At some point in time, 1948 at the latest, the eccentric cams for front- and rear wheel were produced to be identical, with a short threaded end. There is also a through hole for grease lubrication and a grease groove around the eccentric cam spindle.

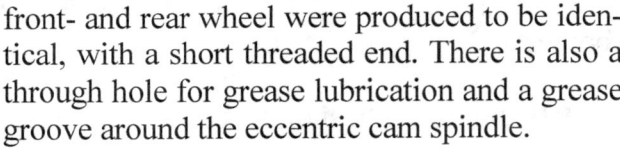

The square hole in the rear brake lever is now placed asymmetrically and marked "S" at one side, so that it can be flipped over and will continue to work correctly.

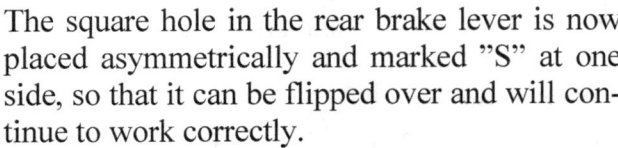

Finally, the last change of the eccentric cam came in the 1950's. The base plate was riveted on, instead of being part of the milled material.

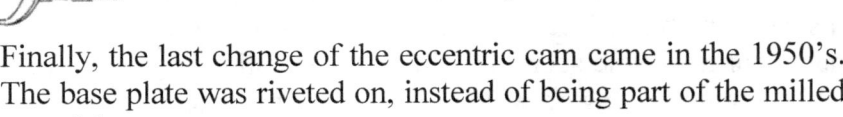

MUDGUARDS

NIMBUS-C mudguards are made from bonderized enameled steel. The mudguards were bought in from various suppliers as raw materials. Therefore variations, such as differences in dimensions of the used steel plate can be found within the same mudguard model.

FRONT MUDGUARD

The mudguards for Nimbus-C comes in two versions, corresponding with the two versions of front fork.
1. Mudguards for the low front fork, fitted rigidly onto the front fork, and the front wheel moves up and down within the mudguard.
2. Mudguards for the high front fork, fitted onto the outer fork tubes, where the mudguard follows the up and down movement of the wheel.

All front mudguards have a rubber mud flap.

Front mudguard 1301 – 1550
Made from 0.8 mm steel sheet with welded-on vallances (sides), also from 0.8 mm steel sheet. Wide sides, round top with angle pieces for fitting the number plate. Note that the mudguards curve follows the wheel from the front fork forward.

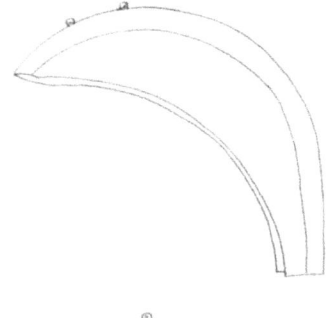

Front mud guard 1551 – 4000
As front mud guard 1301 – 1550 with wide valances and round top, but made from 1.2 mm steel sheet.
The front end from the number plate arches forward, then closely follows the tangent of the front wheels circumference. The welded-on side pieces are made from 0.8 mm steel sheet.

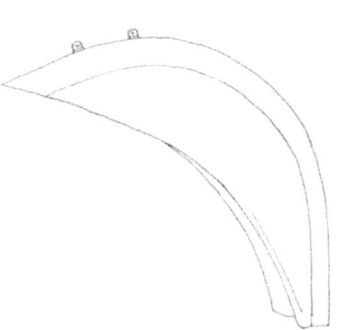

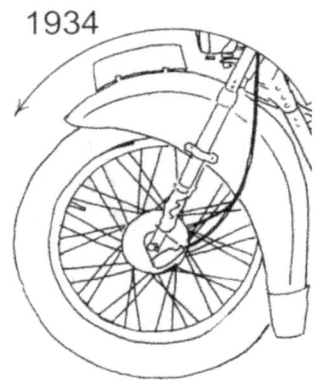 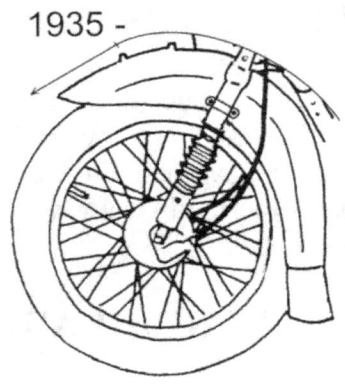

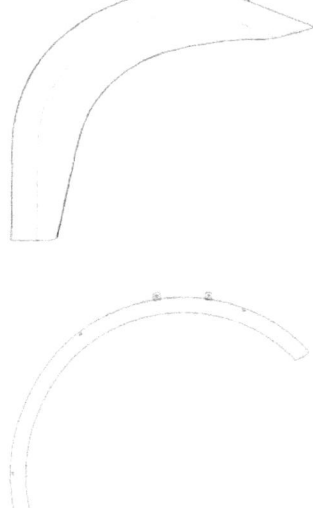

Front mud guard 4000 – 7500 Guard with shallow sides and a round top. The mudguard itself is made from 1.5 mm steel sheet, but the welded-on sides are from 0.9 mm steel sheet, with angle brackets for the number plate. This version was first introduced for 'Sport' versions only and for all machines after 5101 for all versions. A later version, with a beaded edge, and a slightly upwardly bent front tip could be obtained as spare part for machines 1301 – 7500.

Front mudguard 7501 – 8800
Guard without sides and a slightly rounded top from 1 mm steel sheet, with angle brackets for the number plate. The mudguard was so open, that problems with air turbulence affected other parts of the construction (See Chapter Rockers and camshaft housing).

Front mud guard 8801 – 14015
For no obvious reason, during the rest of the production period two types of mudguards were used, both with a slightly rounded top:
One type front mudguard with shallow sides, stamped from 1.2 mm steel sheet.

The other type with the same top section, but fitted with folded-on sides, also from 1.2 mm steel sheet.

The later type mentioned was used from approx. 1950 till March 1956. This front mudguard was specially produced for A/S Fisker & Nielsen by a supplier in England, which means, that there are minor differences in finish and fitting.

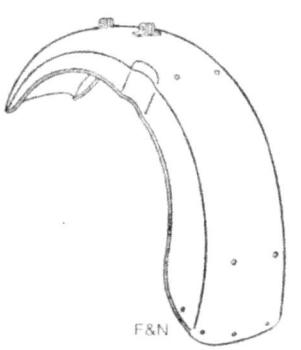

Number plate angle brackets – front mudguard

The front number plate is fitted in a pair of angle brackets made from 2 mm steel, riveted on top of the mudguard between the front end and the front fork. After April 1st, 1956, the front number plate and the brackets were forbidden. That means of course, that all machines after 13573 were supplied without a front number plate and brackets, but this also meant that earlier front number plates and holders had to be removed. A/S Fisker & Nielsen designed a rubber cover shield, to hide the holes in the front mudguard. It is uncertain if these covers were produced at all. The front number plates just needed to be removed; it was not mandatory to turn them in to the authorities. (See chapter Number plates).

REAR MUDGUARD

The rear mudguard for Nimbus-C has basically been unchanged during the whole production period. Two different rear ends catch the eye, whereas a couple of other changes are less clear to see.

Rear mudguard 1301 – 1550

Without brackets for the number plate and with a straight end from the tail light to the bottom rear tip. The top is flat. The sheet thickness was 0.8 mm with a malleable iron hinge machined with recesses on both sides for the pillion seat. There is one hole on either side of the mudguard for the pillion seat spring.

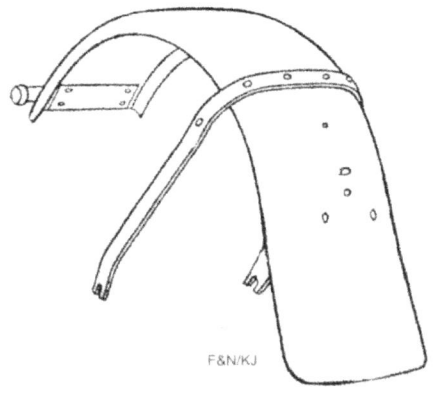

Rear mudguard 1551 – 7500

As 1301 – 1550, but the sheet thickness was increased to 1.2 mm and the hinge is from malleable iron with bulb-shaped ends.

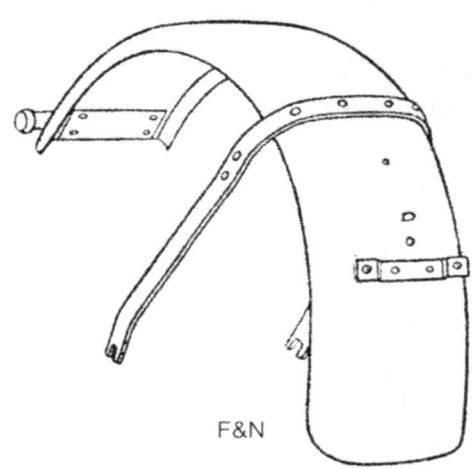

Rear mudguard 7501 – 9000

Sheet thickness 1.2 mm, with brackets for the number plate. The top side is round but the mudguard has a shaped rear tip below the number plate as was the case before 7500.

Rear mudguard 9001 – 12500

As rear mudguard 7501 – 9000. The mudguard braces have two holes at either side for the support of the heavy gauge seat suspension rubber bands. The rear end of the mudguard is curved and the top side is flat as with 1301 – 7500.

SPEEDOMETER

The Nimbus-C can be fitted with either of two speedometers, of differing construction.
1. One construction, working on the principle of a rotating magnet that activates the needle ('VDO')
2. Another construction, where a number of gearwheels with rotating masses activate a clock-like mechanism ('Smiths' chronometric).

With both types, the km-counter is directly driven by the speedometer cable. Both constructions are found in different designs and variations.

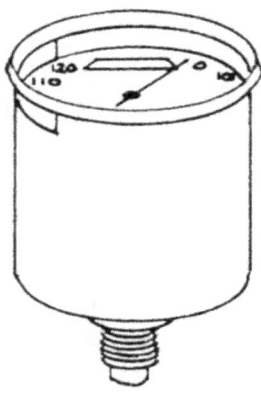

Speedometer 1301 – 2550

'VDO' make, 60 mm diameter. Black dial, white lettering, indicating 0 – 120 km/h with 0 at the '2 o' clock' and 120 at the '10 o' clock' position. External, indirect illumination by means of a lamp, fitted between the speedometer and the ammeter (See chapter Electrics). Ratio 1 : 3, W = 1.5 (One rotation of the front wheel drives the cable 1.5 turns).

Speedometer 2551 – 2900
As 1301 – 2550, but with a bright dial and black lettering, reading 0 - 120 km/h with 0 at the '8 o'clock', and 120 in the '4 o'clock' position. Internal illumination with a lamp, fitted on the base of the speedometer housing.

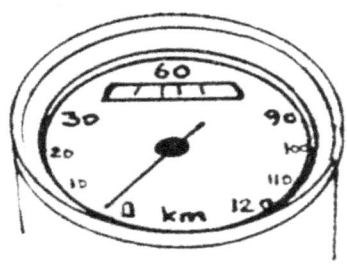

Both mentioned speedometers have their own drive, where a crown wheel on the front hub drives the pinion on the speedometer cable; one rotation of the front wheel corresponds with 1 ½ rotation (W = 1.5) of the inner speedometer cable (See chapter Front wheel).
The union nuts at both ends of the speedometer cable, used with these speedometers, have an internal thread of M17 x 1.5, corresponding at one end with the connection on the speedometer and at the other end with the bushing in the speedometer enclosure on the front wheel.

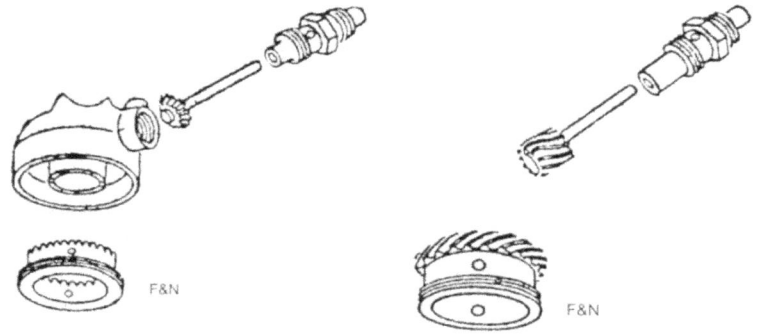

Speedometer 2901 – 7246
'VDO" make, diameter 60 mm. Internal illumination with either a screw fitting or with a fitting pressed into the base of the speedometer housing. Ratio 1 : 2, W = 1.0.
The speedometer drive with a ratio of 1 : 2 is constructed with a worm, driven by a worm wheel. The bushing for the worm has an M18 x 1.5 thread at one end to be screwed onto the speedometer cable and the other end to be screwed into the worm wheel housing in the front hub.

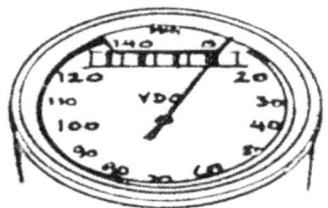

The dial comes in two versions, indicating 0 – 140 km/h, with 0 in the '2 o'clock', and 140 in the '10 o'clock' position with:
   1. a scale that is silk screen printed at the inside of the in-

strument glass; the actual dial disc forms the background with a neutral grey color. There is a gap around the disc for illumination. .
2. a scale that is silk printed with black lettering and scale on a silver-gray disc. On top, between 140 and 0, the disc is partly elevated a few millimeters, to create a gap for illumination.

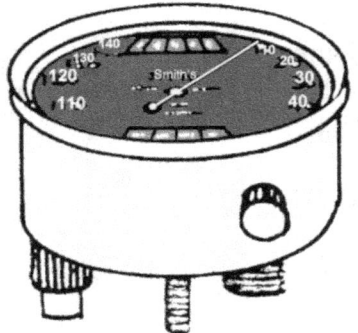

Speedometer 7247 – 9500 (14015)
'Smiths' make, diameter 80 mm. Black dial with white lettering, indicating 0 – 140 km/h with 0 in the '2 o'clock', and 140 in the '10 o'clock' position. Internal illumination. Ratio 1 : 2, (W = 1.0).
There are two versions of 'Smiths' speedometers for Nimbus-C, one without, and one with a trip counter which can be zeroed with a thumbscrew.

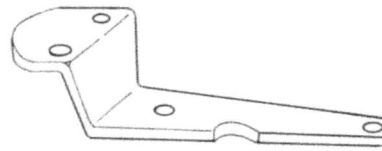

'Smiths' 7247 – 7500
The 'Smiths' speedometer is fitted on a bracket at the mid-front of the handlebars, in the place where the horn was usually fitted.

The horn was fitted, instead, on the frame under the seat on these 253 machines at the place where the guide plate for the hand-change gear lever used to be.

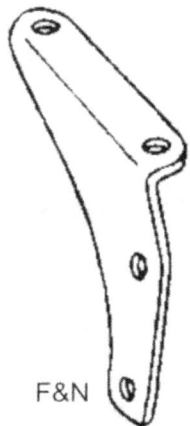

'Smiths' 7501 – 9500 (14015)
The 'Smiths' speedometer is mounted onto an angled bracket, fitted on the right hand headlight bracket.

'Smiths' 9501 -14015
Machines for the military forces were fitted with a 'Smiths' speedometer, fixed on the right hand headlight bracket.
All other machines were fitted with a speedometer of 'VDO' make (see below).
The bushing and worm for the 'Smiths' speedometer and cable are basically as for 2901 – 7246, but both are a little shorter and have a 26 tpi thread on the lower end of the union nut of the speedometer cable.

Speedometer 9501 – 14015

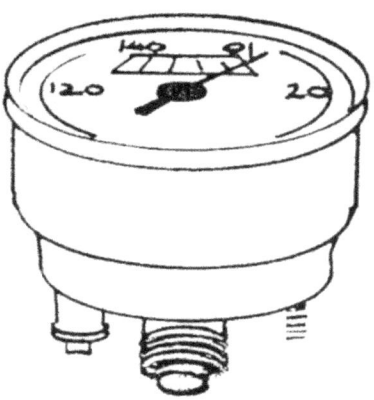

A VDO' make, diameter 80 mm fitted into a "Hella" headlight (See chapter 'Headlight'). Light coloured dial with black lettering and marking, showing 0 -140 km/h with the 0 in the '2 o'clock', and 140 in the '10 o'clock' position. Internal illumination. Ratio 1 : 2, W = 1,0.

The speedometer drive as for 2901 – 7246.

'VDO' speedometers have been found with a diameter of 80 mm, with the threaded connection for the cable at a different angle with the speedometer housing. These are therefore not suitable for fitting into the housing of the 'Hella' headlight.

Other speedometers

Speedometers of different makes with a diameter of 60 or 80 mm and a ratio as mentioned before are sometimes fitted on the Nimbus-C. Examples are 'Weigel', 'NSU' or 'OK'.

ELECTRIC SYSTEM

The electric system of Nimbus-C consists of:

- Battery
- Dynamo
- Voltage regulator and cut-out relay
- Ignition coil and distributor
- Ignition cables and sparking plugs
- Combination switch and charge warning light or ammeter
- Horn and horn button – brake light switch
- Head- and taillight – electric wiring
- External instrument illumination and fuse holder with fuse (where applicable)

BATTERY

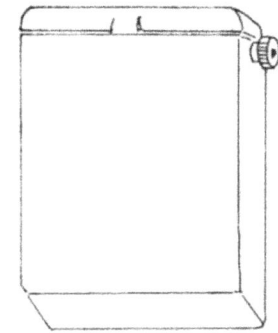

The battery for Nimbus-C can be of different make, but is always a 6 V battery in a black Bakelite housing; the cells are sealed with asphalt. Terminal connections are by external thumb nuts, originally made from brass, later from plastic, red for the positive and black for the negative pole (earth).

DYNAMO

The dynamo for Nimbus-C comes in two different versions:

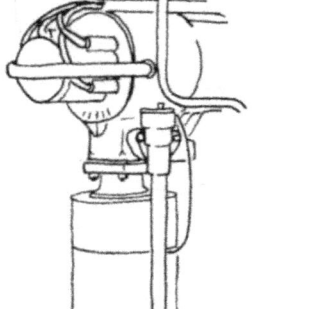

Dynamo '34 1301 – 1550
The first dynamo is constructed in accordance with the third brush principle, which means that the positive pole of the dynamo is connected to the positive pole of the battery, the corresponding negative pole of the dynamo (minus) is connected to the frame via an oil pressure operated cut-out switch (minus). The minus of the battery is directly connected to the frame, through a fuse. The connection between the dynamo and the frame is not established until there is sufficient lube oil pressure to close the cut-out switch. The third (carbon) brush of the dynamo is connected to the field coil and is manually shifted to high or low voltage, and by doing this, the required load current is set.

A half round cover plate on the dynamo has marks from 1 till 4, indicating the load current in Amps. The default setting when starting is 4 and after a long run (with no lights on) a lower value can be chosen. As early as in 1934 the system proved to be unreliable, and was the reason why the factory offered owners a replacement dynamo, free of charge (By the way, at the same time the lube oil system could be converted to the improved version).

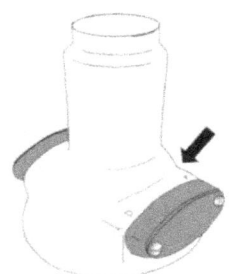

Dynamo '35 1551 – appr. 5000
The second dynamo is a two brush dynamo, with a power of 70 Watt. The loading voltage is controlled by a voltage regulator fitted under the seat. Dynamo front and end brackets are cast in aluminium and polished, and can be recognised by the way it is machined at the front bracket. The area between the 'neck' and the brush holder is rather sharp, giving the impression of a kink. The distance between the fitting plane and the brush holders is 68 mm. The armature of this dynamo is fitted with a commutator with a diameter of 32 mm. The dynamo end bracket is cast with four square projections.

Dynamo '39 appr. 5000 – appr. 10000
Front- and rear bracket are cast in aluminium and polished. The area between the 'neck' and the brush holder is less prominent. The distance between the fitting plane and the brush holders is

increased to 71 mm, which made it possible to increase the diameter of the dynamos commutator to 35 mm. The four cast projections on the rear bracket are now rounded off.

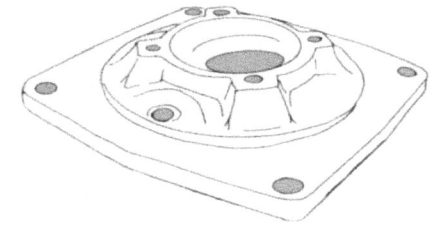

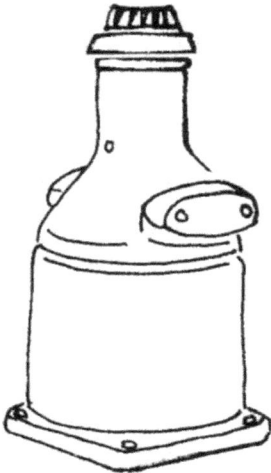

Dynamo '51 appr. 10000 – 14015
The top gear wheel was modified. As with the camshaft gearwheel, the gearwheel with skew teeth became obsolete. Both gearwheels used straight cut teeth instead.

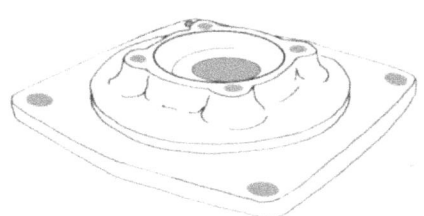

CUT-OUT RELAY AND VOLTAGE REGULATOR

1934 Nimbus-C is fitted with a separate current regulator, via the third brush, and a cut-out relay.
All later versions of the motorcycle have a combined voltage regulator and cut-out relay, in this chapter called the "regulator".

Cut-out relay 1301 – 1550
An oil pressure switch is fitted that acts as a cut out relay. It does this by connecting and disconnecting the dynamo minus pole to and from the frame with changing oil pressure.

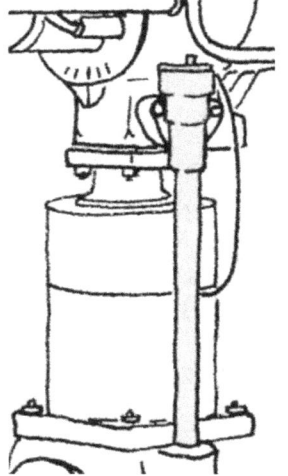

The regulator on the Nimbus-C comes in two main versions: one produced by A/S Fisker & Nielsen and one "Bosch" version.
The factory's own design comes in three different versions.

Regulator '35 1551 – 2050
The regulator consists of a Bakelite plate, with two electromagnetic coils fitted to it making a combined Voltage regulator and a cut-out relay. Both the voltage regulator and the cut-out relay, can be adjusted with screws. The nickel plated brass cov-

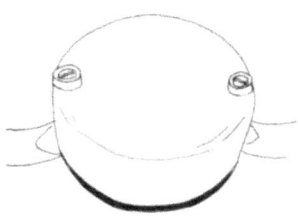

er is kept in place by two elongated brass studs with M4 screws. The head of the screws nestle in a small cup on the cover, where sealing wax was applied to fix the screws. If previously untouched by owners, the screws may be found sealed with sealing wax impressed with "F&N". The voltage regulator and cutout relay is fitted in place with a flat, stiff brass bracket under the seat.

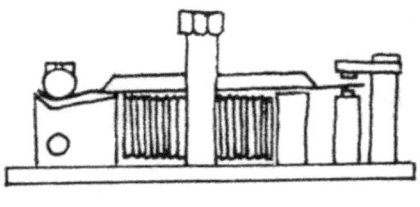

Regulator '36 2015 – 7500
As 1551 – 2050. There is only one brass stud to hold the cover, still by an M4 screw in a small cup, wax sealed and "F&N" marked as before. The attachment bracket is sprung and fitted with two reinforcement pieces.

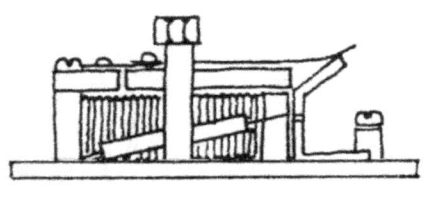

Regulator '48 7501 – 12600
As before 7500, but the two electro magnets do not have sealed setscrews. Adjustment, if needed, must be done by bending the contact plates.

Regulator '54: "Bosch" regulator. 12601 – 14015
The regulator is marked "R.S./T.B./30-45/6.1"
This regulator is fitted with a robust bracket.

IGNITION COIL

The ignition coil for Nimbus-C is combined with the distributor and comes in one model with three versions; all have housing and cover in brown Bakelite.

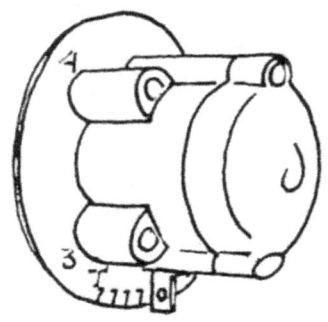

Ignition coil 1301 – 2050
The cover and the housing have the same diameter. The lower part of the cover bowl has cast-in adjustment marks with the characters "S" and "T" for Sent (late) and Tidlig (early) ignition respectively.
The connection with the ignition coil lead is formed by a brass angle piece with an M4 threaded hole.

Ignition coil 2015 – 11300
As before 2050, but the cover now has a collar, encompassing the edge of the housing.

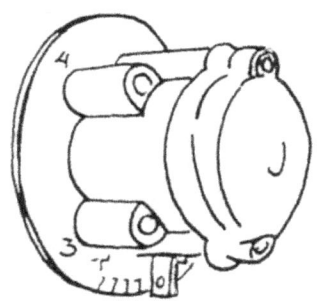

Ignition coil 11301 – 14015
The housing is shaped differently; the area between the ignition cable retainers, coil housing and rear collar are more rounded, instead of sharp. The adjustment marks, and letters below, were deleted and the brass connection for the ignition coil lead is cast into the base of the housing. There is a vent hole in the same place.

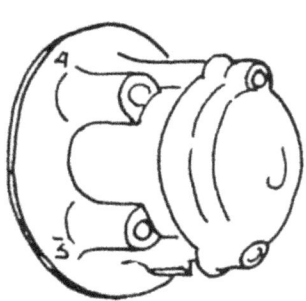

DISTRIBUTOR HOUSING

The Nimbus-C distributor housing, pressed in 1 mm steel plate, is fitted at the front end of the camshaft housing (See chapter Camshaft) and there are two versions:

Distributor housing 1301 – 2050
The distributor housing has two 44 mm long slots for accepting the elongated pins of the bracket for the ignition coils.
From number 1526, the length of the slots were shortened to 20 mm. At the same time, a thumb screw was fitted in the camshaft housing to serve as an earth connection.

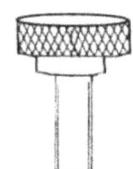

Distributor housing 2051 – 14015
The slots for the pins on the bracket became obsolete and the elongated pins of the bracket along with them. An adjustment plate was riveted at the edge of the distributor housing instead.

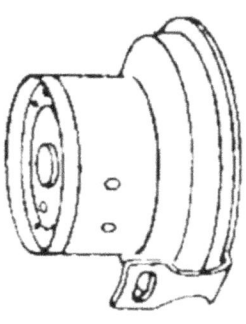

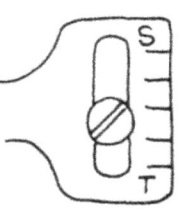

ROTOR

The rotor for NIMBUS-C is of "Bosch" make, during the first year model Z.V.T. 23/1 Z, later Z.V.T. 53 Z1Z and Z.V.T. 53 Z6.

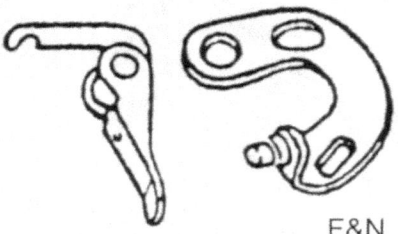

CONTACT BREAKER

The contact breaker for the Nimbus-C can be by different makers, e.g. "Delco Remy" 1855 520, "Doduco" No 301, "Bosch" 1 237 013807 or "Intermotor" 22850.

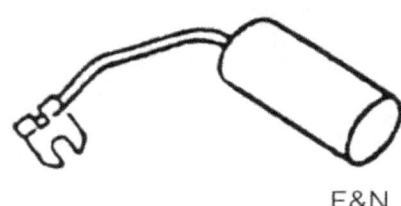

CONDENSER

The condenser for the NIMBUS-C can be by different makers, e.g."Hydra" 5124. The cylindrical condenser is fitted into a nickel – or cadmium plated brass housing, or later made from aluminium,. The condenser is 44 mm long with a diameter of 17 mm and has a capacity of $0.26 - 0.34\ \mu F$ (micro-Farad).

SECURING PLATE

The securing base plate of 0.75 mm steel comes in two versions:

Securing plate 1301 -2050
The securing plate is completely flat and fitted with two brass pins, tightly riveted onto the base of the gearwheel.

Securing plate 2051 – 14015
The securing plate has a Z-shaped form at both ends and has shorter brass pins, tightly riveted onto the base of the gearwheel.

SPRINGS FOR ROTATING MASSES

The wire thickness of the springs is 0.45 mm. During a certain period in 1938, stronger springs with a wire thickness of 0.7 mm have been used. The factory revised this shortly after the change and went back to the previous, weaker springs. Today (2016) some Nimbus owners have chosen to fit the stronger springs, because they claim that the engine runs more smoothly at lower revs.

HT LEADS

All four HT leads on NIMBUS-C have a dedicated soldered-on connection on one end to be connected onto the clamp bushings of the distributor. On the other end, a spark plug cap is fitted. The HT lead consists of a copper lead, encapsulated in insulating rubber. Till 1950, the leads were wrapped in braided linen and covered with black lacquer.

HT leads 1301 – 7500
The four HT leads lay next to each other and are commonly wrapped in isolating linen and reinforced with a fiber stiffener.

HT lead brackets 1301 – 7500
The wrapped HT leads are kept in place by a single bracket of dull nickel plated brass, without being screwed together, fitted onto the foremost stud of the third rocker guide at the exhaust side.

HT leads 7501 – 14015
The HT leads lay next to each other on top of the camshaft housing, fitted with two brackets.

HT lead bracket 7501 – 13572
The HT leads are kept in place by two brackets; one for all four leads, fitted onto the foremost screw of the first rocker guide at the left hand side, and one for two HT leads, viz. the 3rd and 4th cylinder onto the foremost screw of the third rocker guide at the left hand side.

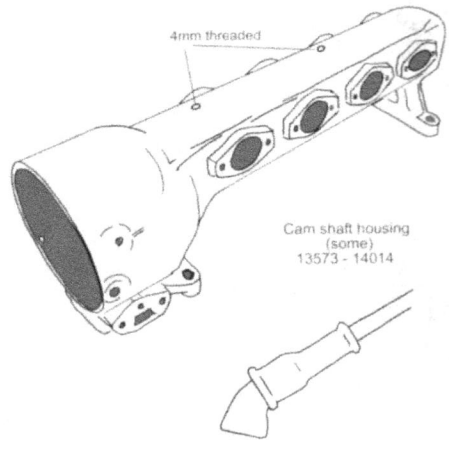

HT lead bracket and rubber spark plug caps applicable to some machines in the range of 13573 – 14015
The short brackets for the HT leads are fitted onto the other side of the camshaft housing with two M4 screws. Fiber discs are fitted between the brackets and the camshaft housing. At the same time, flexible rubber boots were fitted partly over the rigid spark plug caps.

SPARK PLUG CAPS

All original spark plug caps for Nimbus-C are made from brown Bakelite and are the same, but there are various different brass connectors.

Spark plug cap connector 1301 – appr. 7500
The connector bushing is cut into four ribbed blade springs, which are pushed down over the threaded end of the spark plug.

Spark plug connector appr. 7501 – 14015
The bushing is provided with a through-going 1.0 mm spring wire.

SPARK PLUGS

Spark plugs for NIMBUS-C can be of various make. The thread is M14 x 1.25mm, thread length 12.7 mm and the gap is 0.7 mm. The original spark plug was a "Lodge" HD 14. Later, the factory advised a number of other spark plugs, of which codes and some of their names are outdated now.
Today (2016) the following spark plugs are recommended:
"Autolite" 283 (resistor spark plug for built-in ignition noise suppression), "Bosch" W7AC, "Champion" L87 Y, "Champion" L87 YC, "Denso" W20FP-U, "Eyquem" 750, "NGK" B6 HS, "NGK" BP6 HS or equivalent.

COMBINATION SWITCH

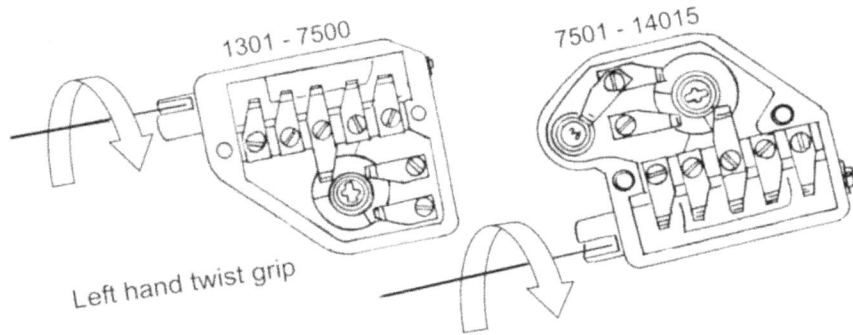

At the underside of the steel plate handle bars, Nimbus-C is fitted with a combination switch, fitted in a Nokait/Bakelite housing. It has an ebonite roller, a brass lead, a Nokait/Bakelite ignition lock with a brass switch plate. The combination switch comes in two versions:

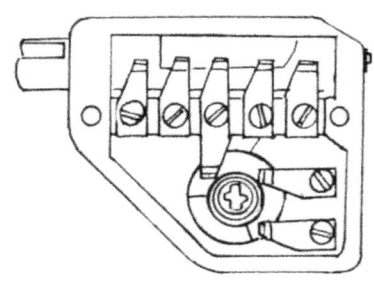

Combination switch 1301 – 7500
The ignition key switch has three positions: Off, Park and Run. For 1310 – appr. 6400, these positions are indicated with the letters A (Afbrudt = Off), P (Parkering = Park) and K (Kørsel = Run) on the ignition key surround. The switch drum has four positions: 1 OFF, 2 Parking, 3 Low beam, 4 High beam.

The combination switch for machines 1301 – 1550 have a brown Bakelite ignition key switch of the same material the ignition coil and the fuse holder are made from. After 1551, the controller and the ignition key switch are both made from black Nokait (Bakelite).

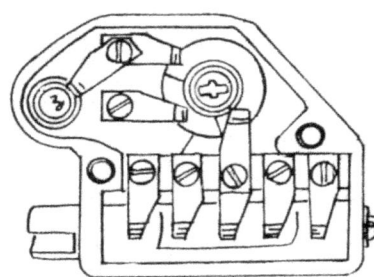

Combination switch 7501 – 14015
The basic principle is as before 7501, but it is a mirror image of the previous construction. It has with a built-in charge warning light. In addition, the text on the bottom side is also in English.
There are two versions, one with and one without a collar around terminal D of the charge warning light.

The glass for the charge warning light on the handle bars is part of the combination switch.

IGNITION KEY

All ignition keys for Nimbus-C are identical!
The key is stamped from 2 mm brass plate and matt nickel plated.

AMMETER

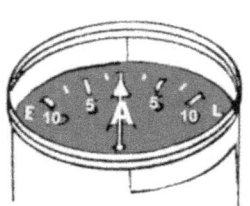

The ammeter for Nimbus-C is fitted on the left hand side of the handle bars, Ø 52 mm.

Ammeter 1301 – 1550
"Schoeller" make. Black background with white lettering. Maximum scale reading + / - 6 Amp. The + is marked "L" (Lade = Charging) and the – with "E" (Entladen = Discharging).

Ammeter 1551 – 2400
"VDO" make, reading + / - 10 Amp.; for the rest as 1301 – 1550.

CHARGE WARNING LIGHT

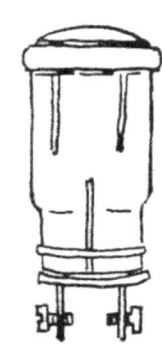

Charge warning light 2401 – 7500
"Bosch" make, model JJ 7/2 with red glass, fitted in make- and number plate on the handle bars.

Charge warning light 7501 – 14015
"Philips" make, model 6876, built in into controller.

HORN

Horns from at least three different make were fitted onto the Nimbus-C, and maybe more.

Horn 1301 – 7500
"Riemann" make. Distinguishing characteristic: crossed oak leaves, stamped into the front plate. The horn is fitted onto a special bracket on the front bottom of the handle bars. Bakelite muffs are fitted around the jacks of the Riemann horn.

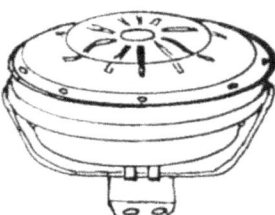

Horn 7501 – appr. 9000
"Klaxon" make. Distinguishing characteristic: A brass shield with the manufacturers name and number. Fitted with a stiff angled bracket onto the yoke of the front fork.

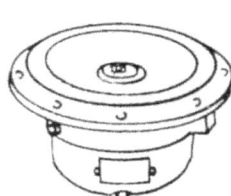

Horn appr. 9000 – 14015
"Hella" make. Distinguishing characteristic: The manufacturers name plate is screwed onto the front plate or moulded into the rubber cover of the front plate. Fitted onto the yoke of the front fork by means of a sprung bracket.

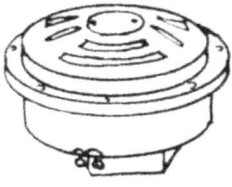

PUSH BUTTON FOR THE HORN

Push button 1301 – appr.5400
"Bosch" make, model SSH 506/1z with screw-on top, made from aluminium. Fitted onto the handle bars at the left hand side.

Push button appr. 5401 – 7500
As 1301 – 7500, but closed by means of a spring. Chrome plated brass, aluminium and zinc have been used as materials.

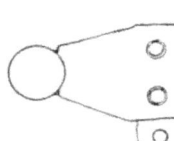

Push button 7501 – 14015
Fitted onto the left hand side of the handle bars. Earth connection through twist grip.

BRAKE LIGHT SWITCH

The brake light switch on Nimbus-C comes in two versions, both cast in black Bakelite.

In 1934, brake lights were not mandatory, so the first motorcycles most probably hit the market without being fitted with brake light and –switch.

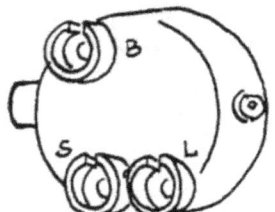

Brake light switch 1301 – 1550
The construction of the housing is such that the distance between the frame and the cast connection in the housing for the threaded pull rod is 7 mm.

Brake light switch 1551 – 7500
The construction of the housing is changed; the distance between the frame and the cast connection in the housing for the threaded rod is decreased to 2 mm.
The cast connection to the threaded pull rod is now reinforced.

Brake light switch 7501 – 14015
The threaded pull rod now has an integrated 'head'.

HEADLIGHT

The headlight for Nimbus-C comes in two different basic designs, related to the two types of front forks:
- One with single point fixing for the low front fork
- One side fitted for the high front fork.

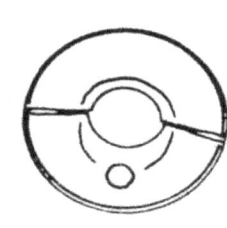

Headlight 1301 – 3000
"Riemann" make with a single parabola. The headlight glass has vertical longitudinal ribs and is slightly convex. The "Riemann" logo, two crossed oak leaves, is stamped at the rear of the lamp housing together with the model number.

Head light 3001 – 7064
As 1301 – 3000, but with a kinked parabola, a parabola with different focal points, one for low beam and the other one for high beam. The headlight glass has vertical longitudinal ribs and is flat.

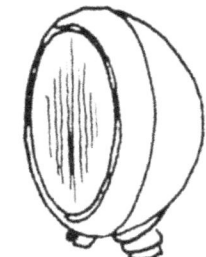

Head light 7065 -7500
"Lucas" make, model DU 42 with single point fixing. After 1949 the "Hella" headlight with single point fixing was available as a possible substitute for the headlight for 1301 – 7500.

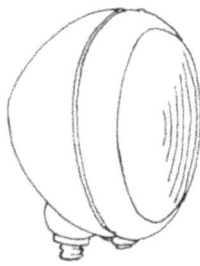   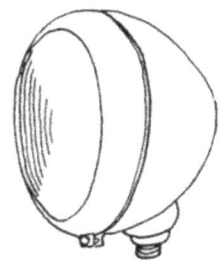

Lucas DU 42   *Hella*

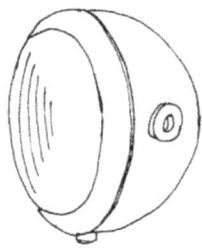

Headlight 7501 – appr. 8000
"Lucas" make, model MU 42, two-hole side fixing.

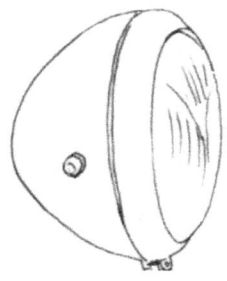

Headlight appr. 8000 – 9500
"Hella" make two-hole side fixing.

Headlight 9501 – 14015
"Hella" make, side fitted with a hole for the speedometer (See chapter Speedometer). Hella parabola with parking light for a 7 inch headlight: 1A6 002395-191

or ...

"Lucas" make, side fitted. Diameter front opening 120 mm, as opposed to 150 – 160 mm for all other headlights. Named FKF STD 23-609.1-2. Fitted onto machines for the military forces.

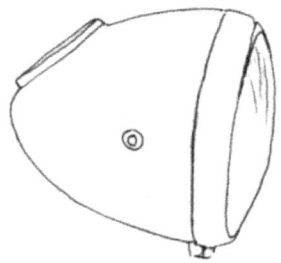 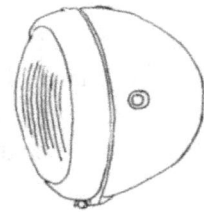

TAIL LIGHT

The tail light for Nimbus-C has a cylindrical shape, made from aluminium with sockets for two light bulbs. The glass is fitted in a groove and is kept in place by a wire spring. Perhaps a more simple taillight was fitted before the brake light became mandatory in 1934. The taillight comes in two versions:

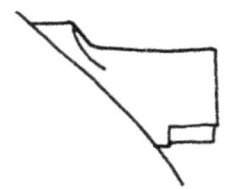 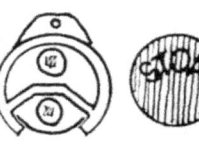 

Taillight 1301 – 1550
The tail light housing is fitted without a gasket, directly onto the rear mud guard. It has an identifying 'bulge' on top. In the inside, a separator is cast between the tail light and the brake light partition. A 6 mm hole for the

wires is drilled at both sides of the housing. The tail light functions also as a number plate light, by means of a slit covered by a celluloid plate at the underside.

Taillight 1551 – 2050
As 1301 – 1550, but without holes in the side walls. The wires run through a hole in the rear mud guard, directly into the tail light housing.

Taillight 2051 -14015
The shape of this tail light was changed. The top side has a rounded, hollow shape, and in the inside there are two 'keyways' for fitting (sliding) a non-transparent separator into.

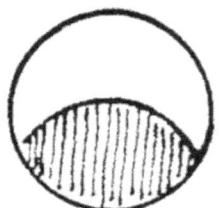

Taillight glass 1400 – 2050
The tail light glass has a red, 0.5 mm thick celluloid plate with the text "STOP". The text is silk printed in mirror image at the top inside of the celluloid plate. The text cannot be read until the brake light is activated.

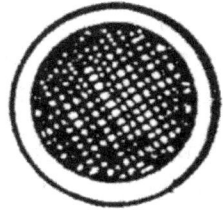
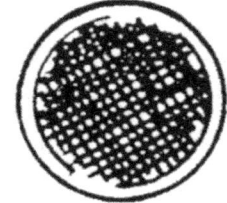

Taillight glass 2051 – 13200
Clear, 0.5 mm thick celluloid plate, divided in a lower red field, for the tail light, and an upper yellow field, for the brake light.

Taillight glass 13201 – appr. 13600
Red reflector ('cat's eye'), is cast in mineral glass and fitted in a chrome plated ring.

Taillight glass appr. 13600 – 14015
Red, plastic reflector marked "J.R.U.129" (J.R.U. = Justitsministeriets Refleks Udvalg, Ministry of Justice Reflector Commision)

Rubber moulded gasket 2051 – 12600

A thin, black, rubber sheet with a thicker edge around.

Rubber moulded gasket 12601 – 14015
A thick, grey rubber sheet, complying with the requirements of the authorities to allow illumination of the number plate.

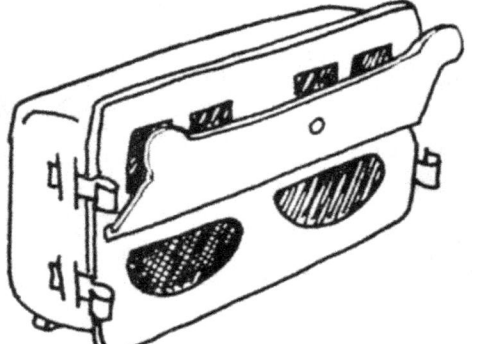

Tail light (military forces) From about 1949, military machines were no longer fitted with the Nimbus tail light, but with a special tail light instead, called "FKF STD 23-609.1-10" made by the German company "Notek".

The flap on the side facing backwards can either (in war time) be flipped down, to cover the two lower glasses, yellow for brake light and red for taillight, and then allow a weak greenish light to be seen, or be flipped up to cover the four upper glasses for normal running, during the day or night.

When the flap is down (night run during war time), two green lights can be seen from a distance. When one has approached so close, that all four lights can be seen, one is too close! At the lower side, there is just a little hole for the brake light. When the flap is down, the white number plate light can (and must) be covered.
Later (after 1956) versions of this taillight have red glasses fitted for rearward shining lights.

Other tail lights
Many machines had a number plate holder with an English "Britax" tail light fitted, upon request by the dealers.

THE ELECTRICAL WIRING

The electrical wiring for Nimbus-C is single wired with 1.5 mm² rubber insulated stranded copper wire, braided with linen and lacquered black. The dual tail light however, is wired with 1 mm² copper wire. For the rest, the wiring is in accordance with the wiring diagram. Single wires are combined in black lacquered linen braiding, kept together with linen insulation tape. From about 1954, the wires are covered with flexible black PVC, kept together with black PVC insulation tape. The wire ends can either be without marking and have a soldered eye at the end, or have a cable eyelet terminal with a stamped number or letter. The markings on the cable terminals are shown in the wiring diagram. Please note that the numbers used on the terminals for 1301 – 1550 are in *italic*.

Wiring diagram 1301 – 1550
Wires for the brake light were added from 1400 on.

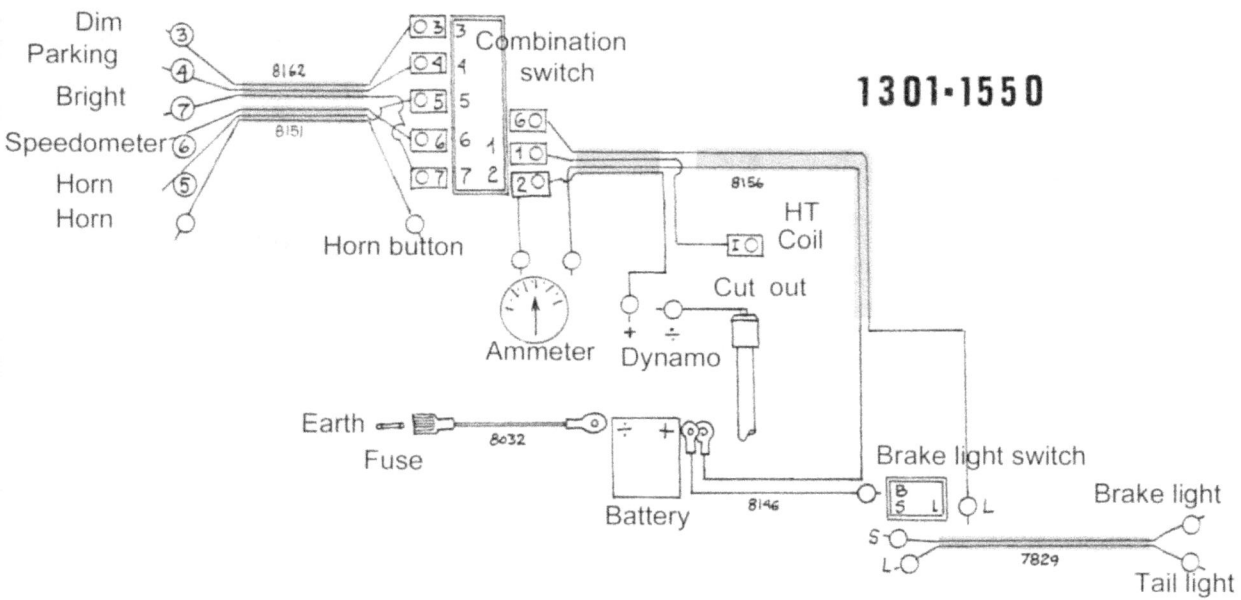

Wiring diagram 1551 – 2400

The dynamo was changed and a combined voltage regulator and cut out switch was introduced.
Wiring diagram 2400 – 7500

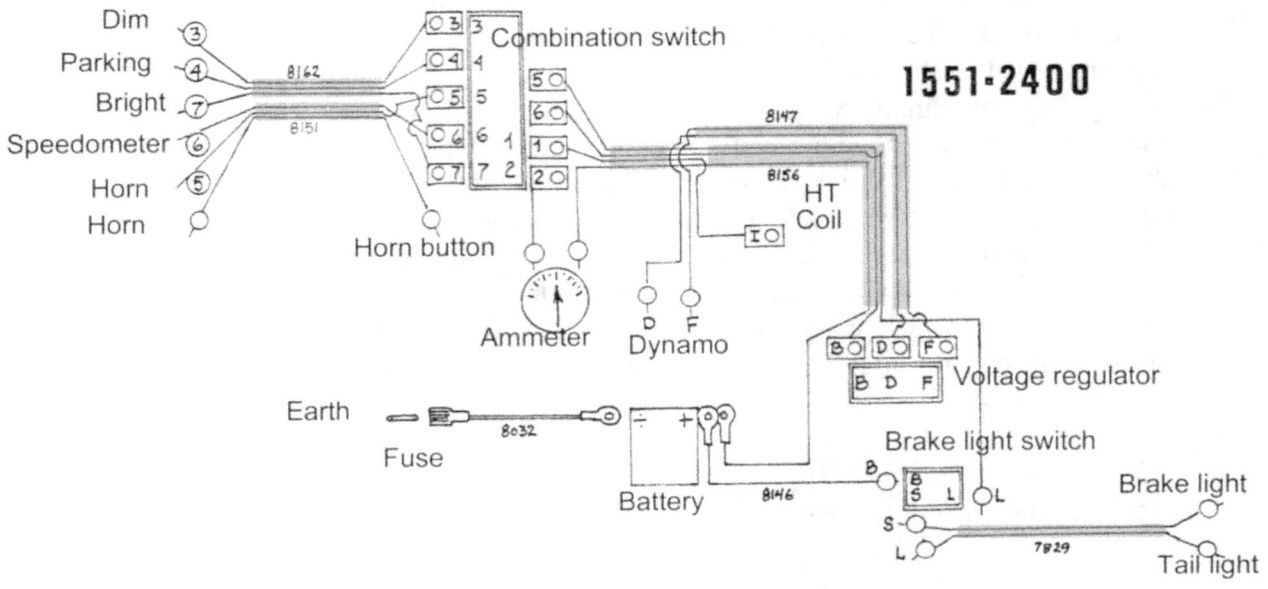

The ammeter was replaced by the charge warning light, resulting in a changed wiring diagram.
Wiring diagram 7501 – 14015

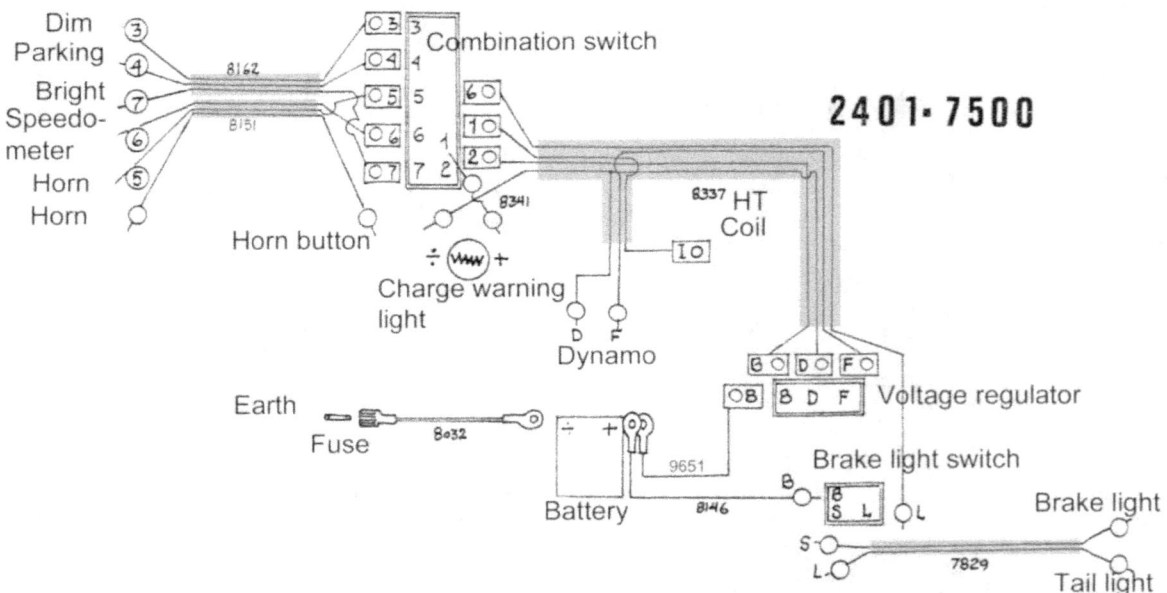

The charge warning light was integrated into the combination switch at the underside of the handle bars, which also resulted in a slightly modified wiring diagram.
Later wiring

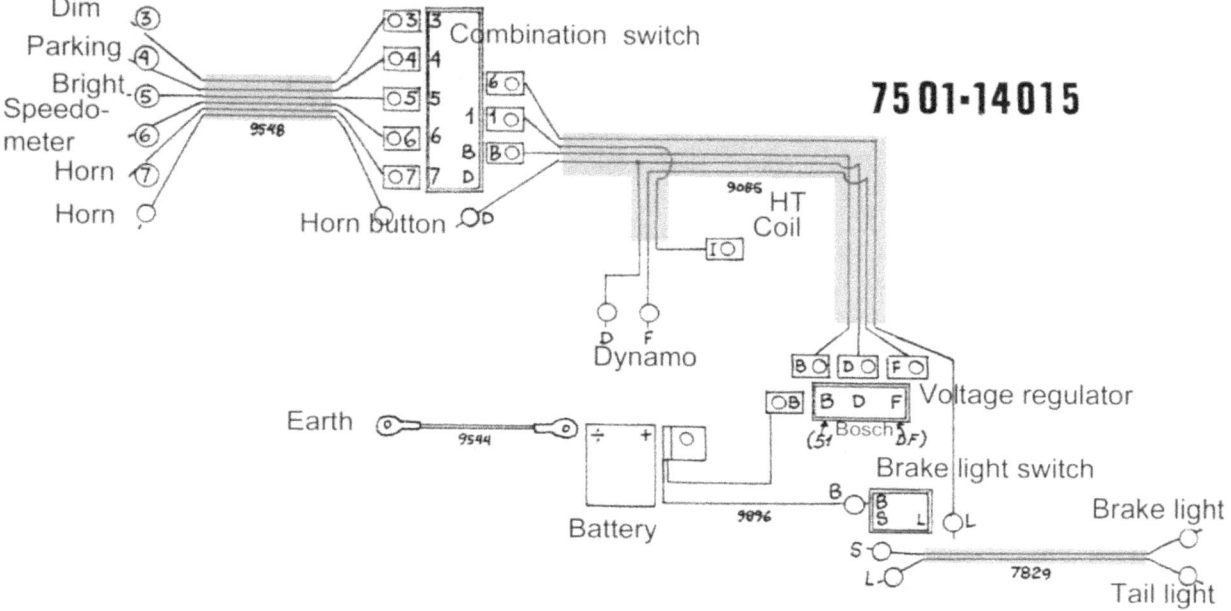

The extruded PVC insulation of the electrical wiring is now in different colors.

FUSE

Fuse holder 1301 – 7500
All machines till 7500 are overload protected by means of a 20 amp fuse. The brown Bakelite fuse holder is fitted in the earth lead that runs from the battery minus pole to a threaded stud on the gearbox.

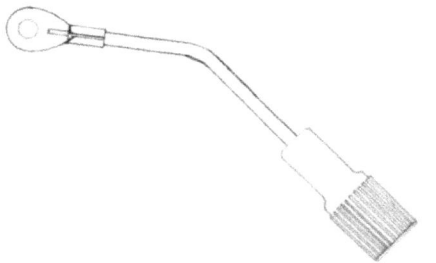

INSTRUMENT ILLUMINATION

Instrument illumination 1301 – 2550

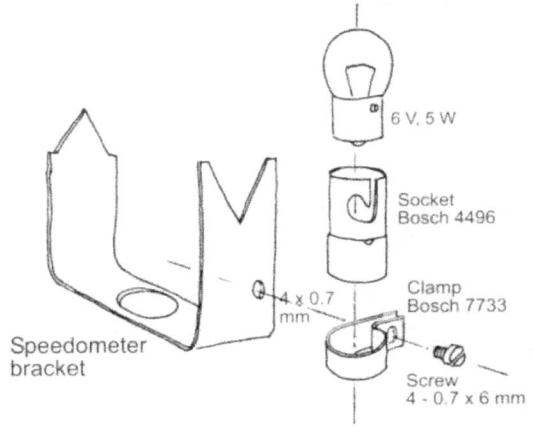

The light for external illumination, for both the ammeter and speedometer, is fitted on the speedometer bracket at the bottom side of the handle bars. The socket (spare part number 4496) is from "Bosch" with a Bakelite base. The clamp (spare part number 7733) is also of "Bosch" make.

Instrument illumination 2551 – 14015.
The instrument illumination light is built into the various types of speedometer.

CABLE CLAMPS, ETC.

Clamp (4572) 1301 – 1550 and 2401 – 14015
Large one-screw bracket. The wire bundle is secured under the frame plate between fuel tank and seat and is fitted with a countersunk M4 screw.

Clamp (4572) 1310 – 7500
Large one-screw bracket. The wire bundle is secured at the lower flange of the steering yoke frame 'shield'.

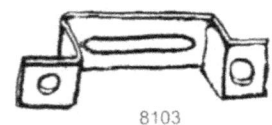

Clamp (4571) 13109 – 14015 Small one-screw clamp. Fitted at three places onto all machines for the twin wires to the taillight on the right hand side on the inside of the rear mud guard.

Flat Clamp (8103) 1551 – 2400
Two-screw bracket. The two wire bundles (8147 and 8156, see wiring diagram) are secured under the frame plate between fuel tank and seat.. Fitted with countersunk M4 screws.

Tightening strap (8197) 7501 – 13572

The wiring is secured by means of an aluminium tightening strap, right under the lower left hand flange of the steering yoke, near the frame reinforcement strip.

Double lead clamp (10947) 13573 – 14015

From frame number 15001, matching with machine 13573 the wire bundle is kept up together with the throttle- and the clutch cable between steering yoke and frame. The single screw clamp is fitted at the inside of the triangular part between the flanges of the yoke. A 6 mm long screw holds the two plates of the steering damper together; between those plates, the lead- and cable clamp is fitted.

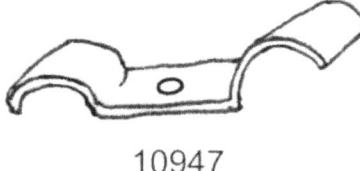

10947

Cable tag (9881) 9001 – 14015
Two lead cable eyelet terminal for connecting the battery + pole with the regulator and brake light switch.

Cable eyelet (7490) 1301 – 140145
Small one-lead cable eyelet terminal, was used for nearly all wires in the wiring loom on all machines. Their marking is shown on the wiring diagrams. For connecting the "Bosch" controller (as of machine 12601) the hole of the cable eyelet was advised to be opened up to 5.5 mm and the remaining metal cut- or filed out to make a spade terminal. Then to be fitted with a screw to connections D+, DF and 51 of the regulator. Tinned finish.

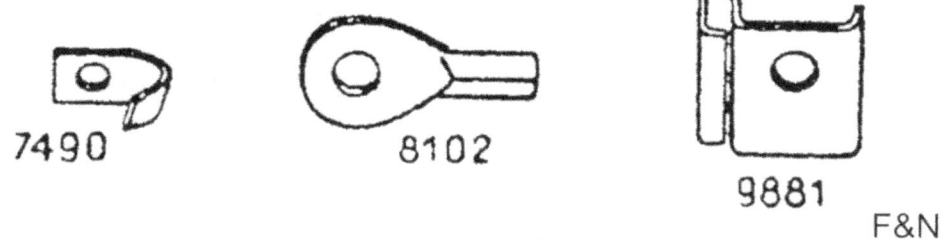

7490   8102   9881   F&N

ACCESSORIES

Nimbus-C was supplied with the following accessories:

- Tool set
- Tyre pump

In addition, the factory offered the following extra accessories:

- Pillion seat with foot rests
- Leg protectors with brackets
- Mud protectors (1935 – 37)

For the military forces were fitted special accessories:

- Head light
- Taillight
- Jiffy stand
- Side bags
  Bumper-to-bumper lights (convoy lights)

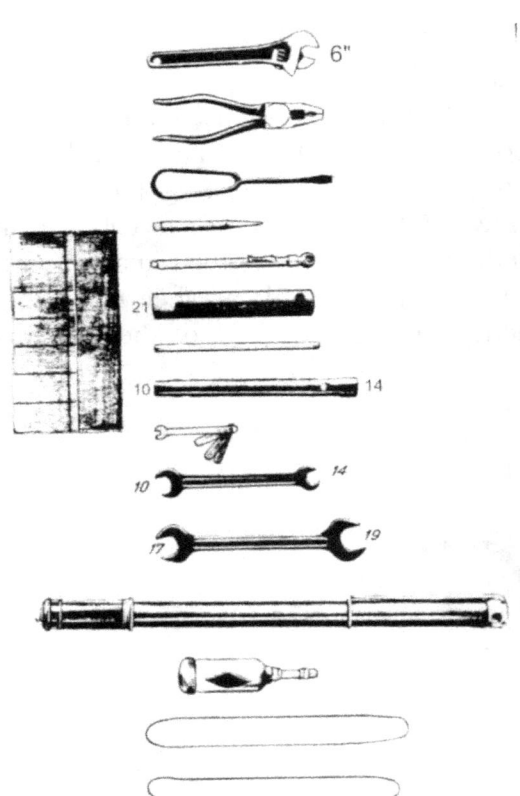

TOOL SET

A tool set consists of

a linen tool pouch containing:
- 6 mm open end spanner with feeler gauges
  (0.3 mm, 0.5 mm and 0.7 mm)
- 7" comination pliers
- Screwdriver
- Spark plug spanner
- Double ended pipe wrench (10 mm and 14 mm)
- Rod for spark plug spanner and two sided pipe wrench
- Two sided open end spanner (10 mm and 14 mm)
  may show the NIMBUS trade name
- Two tyre levers

The tool set for Nimbus motorcycles for the military forces additionally contained:
- Adjustable spanner "Bahco" 8"
- 3 mm mandrel (for driving out split pins)
- Tyre pressure gauge
- Grease gun

The tool pouch for the military forces was made of linen, later plastic foil.

Service tools, with tool list and engine repair stand, was supplied to authorized dealers and military workshops only and not to private persons.

TYRE PUMP

The tyre pump is a hand pump, marked "Bluemel's MOTO-BIKE". The pump comes in various lengths and models. Long pumps are for machines before 9000, shorter ones for later machines. The long tyre pumps are normally fitted with body and hand grip from rolled steel sheet. Short pumps may have a body and / or a hand grip from rolled steel sheet and / or Bakelite. Newer replica's come in a plastic version.

Tyre pump catch 1301 – 9000
The pump is secured at the left hand side under the upper frame section, by means of two hook-shaped catches. The foremost catch is fitted under the nut for attaching the seat, the rear one with a screw, Metric Thread Fine Pitch 8 x 1.0 - 12 mm, fitted into a threaded hole in the frame.

Tyre pump catch and –pin 9001 -14015

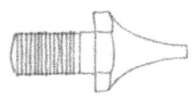

The tyre pump is fitted at the left hand side under the upper frame section by means of a hook shaped catch and a stud. The foremost catch is fitted under the nut of the seat support, at the rear, a special stud is fitted and secured by a nut, Metric Thread Fine Pitch 6 x 0.75.

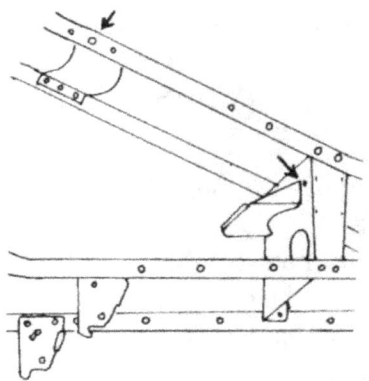

OTHER ACCESSORIES

Pillion seat with foot rests
Pillion seat (See chapter Seat *and pillion seat*); foot rests (See chapter *Frame- Imitation foot rests*)

Leg (weather) protectors with brackets
The leg protectors are made from strong grey-green linen with reinforcements in knees and crotch. In the side hems, there is a bracket from thick steel wire. Below, at both sides, there are hems for bent flat steel brackets, fitted onto the frame by means of the two front engine support bolts.

Mud protectors
In 1935 – 37, the factory sold metal mud protectors for DKK 15.- a pair. These were probably protectors to be fitted at both sides in front of the rider's foot rests. Unfortunately,
these mud protectors are only known from a few photographs.

Equipment for machines for the military forces after 1949

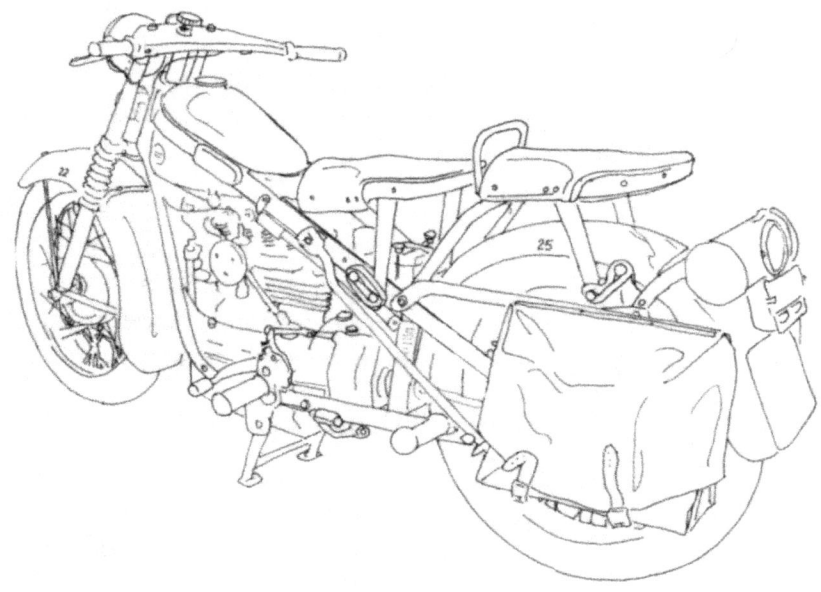

The head-
and taillight for military machines are discussed under *Electrical equipment*.

- The jiffy stand comes in two, slightly different versions, called "model 1" and "model 2".
- Side bags and bumper-to-bumper (convoy) lights also belong to the military equipment.

Other accessories
The Nimbus-C can be fitted with additional accessories such as: searchlight on the handle bars, many other types of jiffy stands, other types of taillight, crash bars, side mirrors, wind screen, side bags, handlebar lugs, and luggage carrier, but are not further dealt with (in this book).

SIDE CARS

Many Nimbus-C's were supplied with a sidecar. A stiff steel construction with an unsprung rear wheel is very suitable for having a sidecar fitted. A/S Fisker & Nielsen produced sidecars from 1920 - 1925 for the Nimbus A/B, and from 1934 – 1959 for Nimbus-C. A Nimbus sidecar could also be fitted on other brands of motorcycles. It is unknown if this happened and if so, to what extent.

Nimbus-C motorcycles can be fitted with other brands of sidecars; insofar they comply with the requirements with respect to testing and registration. The description of these sidecars is beyond the scope of this book.

MODEL AND VERSION

The factory showed a systematic lack of consistency where it comes to the model names for sidecars and their bodies. Therefore we will use the word *version* to describe the side cars chassis:

Chassis number:	Year:	From:	To:
	1934	501 -	550
	1935	551 -	565
		U 701 -	U 814
	1936	U 815 -	U 948
	1937	U 949 -	U 1025
		R 1201 -	R 1240
	1938	R 1241 -	R 1403
	1939	R 1404 -	R/RB 1800
	1940 - 44	R/RB 1801 -	R/RB 2212
	1945	R/RB 2213 -	R/RB 2288
	1946	R/RB 2289 -	R/RB 2425
	1947	R/RB 2426 -	R/RB 2618
	1948	RB 2619 -	RB 2827
	1949	RB 2828 -	RB 3130
	1950	RB 3131 -	RB 3621

Chassis number:	Year:	From:	To:
	1951	RB 3622 -	RB 4099
	1952	RB 4100 -	RB 4744
	1953	RB 4745 -	RB 5106
	1954	RB 5107 -	RB 5507
	1955	RB 5508 -	RB 5768
	1956	RB 5769 -	RB 5876
	1957	RB 5877 -	RB 6015
	1958	RB 6016 -	RB 6040
	1959	RB 6041	RB 6194

Version 34: Tube chassis without brake, version '34 (chassis no. 501 – 565)

Version U: Flat steel chassis without brake, version '35 (chassis no. U 701 – U 1025)

Version R: Tube steel chassis with brake, version '38 (chassis no. R 1201 – R 2540)

Version RB: Tube steel chassis with brake, version '38 (chassis no. RB 1201 – RB 2540)

Tube steel chassis with brake, version '48 (chassis no. RB 2541 – RB 3424)

Tube steel chassis with brake, version '50 (chassis no. RB 3425 – RB 6194)

The term *version* is used in this book for various sidecar constructions. The factory used the term *model* for years, chassis and construction. This may be confusing. The passenger sidecars were called *Model 'Standard'*, *Model 'Luxus' / Luksus* and *Model 'Sport'* and for pick-up sidecars the name *Model 'Standard'* was used. A special version from 1939 was called *Ambulance*.

In the following section are dealt with.
- sidecar chassis, spring suspension and attachment to the motorcycle
- sidecar wheel, mud guard and light

Passenger- and pick-up bodies will only be briefly described in this book.

CHASSIS NUMBER

The first sidecar number issued for a Nimbus-C is 501. Preceding numbers were issued for Nimbus A/B (Stove pipe). Up until 1956, the sidecar number for a Nimbus was stamped on a metal plate, riveted onto the chassis.

On a chassis with semi elliptical springs, (version R), the plate bearing the serial number is riveted onto the clamping plate, where the springs are attached to the chassis, on the side nearest the motorcycle.

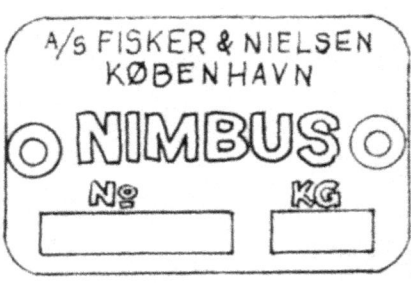

A chassis with quarter elliptical springs (model RB), the chassis number plate is fitted on the tube chassis between the two clamping brackets on the side nearest the motorcycle.

After April 1st, 1956, the chassis number was stamped onto the tube chassis, reading *NIMBUS-RB* followed by the chassis number, on the side of the rear clamping stud.

CHASSIS

There are several sidecar chassis versions for Nimbus-C. The majority have a chassis frame of 600 mm wide, while the distance between the front and the rear body hinge varies with the suspension used.

Some chassis frames, constructed for special applications, deviate from the standard dimension of 600 mm and are 675 mm wide.

All chassis frames for sidecars are black enameled.

Three of the chassis versions have a tubular frame:
- One without brake (1934)
- Two with brake (Version '38 and version '48 / '50)

From 1935 –'38 a flat steel chassis without brake was produced (Version '35)

Tubular chassis without brake
Chassis version '34, no. 501 – 565 (Version '34)
Tubular chassis with four ball-and-socket connectors for the sidecar to be coupled to the motorcycle, two on the lower end and two through two angled rods at the higher end.
The side car body has two transverse wooden beams fitted at the front under the base. The beams have two pairs of clamping brackets with leather or rubber dampers fitted around the tubes of the chassis.
The side car body is prevented from moving sideways over the chassis by the clamping brackets, which are secured by two steel stop rings welded onto the tubes of the chassis.
At the rear bottom side of the body, the suspension with two transverse pull bars is fitted. Between those, there are two studs for two strong expansion springs, which are connected to the rear of the chassis frame.
The construction using this type of springing was inadequate and very susceptible to the body striking the chassis

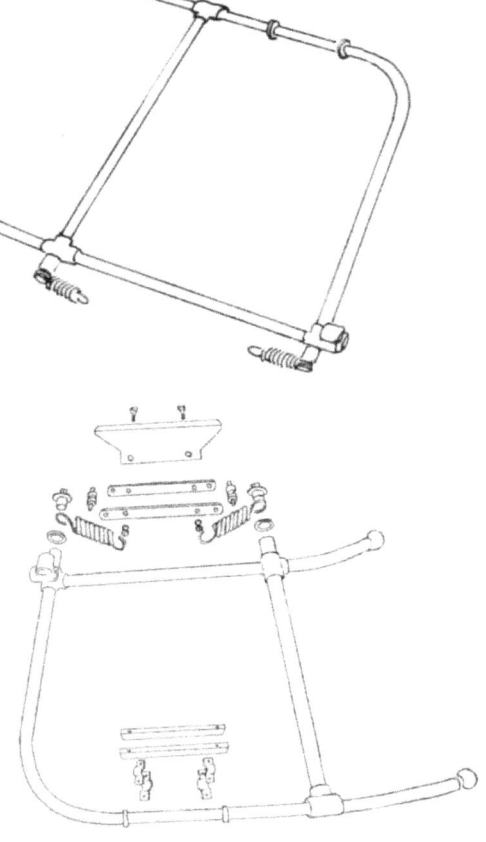

Flat steel chassis without brake:
Chassis, version U, no. 701 – 1025 (version '35)
Flat steel chassis, which is logical in relation to a motorcycle with a flat steel frame. In practice , the chassis regrettably appeared to be so flexible, that it was called 'floppy'.
The chassis has four ball-and-socket connectors. The two angled rods are identical with balls at both ends. The wheel has no brake. At the front, the body is fitted on two rubber blocks on a steel side member. At the rear, the body is sprung by heavy gauge rubber bands or by two compression springs. The rubber bands come in two versions, one for a passenger body and a more robust version for a pick-up body, while the compression springs are the same for both versions.

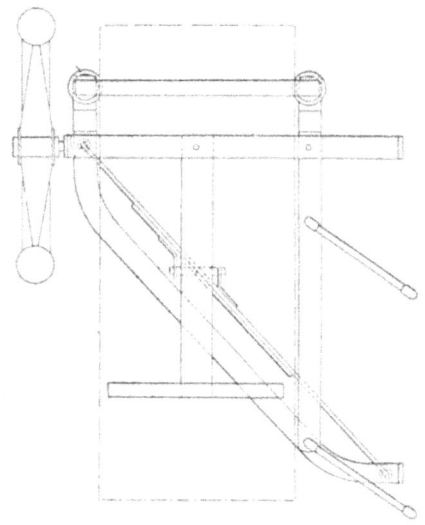

Tubular chassis with brake:
The springing of the tube chassis comes in two versions:

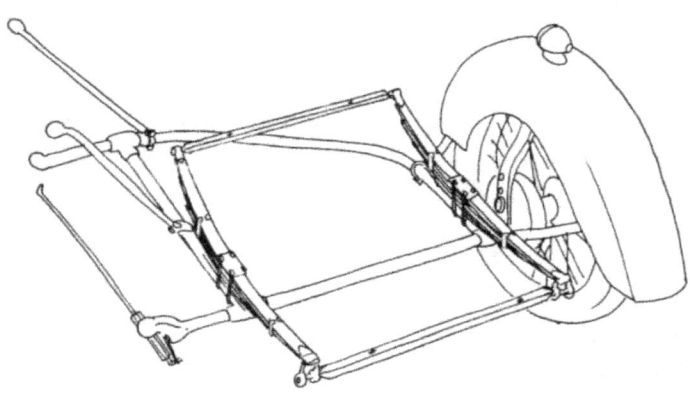

Version R, with semi elliptical leaf springs. On the chassis frame, two sets of longitudinal blade springs are fitted, that look like the bottom halve of an ellipse.

For a passenger body, the leaf spring assembly has five steel blades, for a pick-up body eight. At the front- and rear ends of the blade springs, a pivot is fitted. The front is connected to a cross bar, and at the rear to a pair of links connected to a rear cross bar.

The links come in three versions, short, medium and long.

For weak springs with five steel blades and / or for a passenger body, short or medium links are used. For strong springs with eight steel blades and / or for a pick-up box, medium or long links are used.

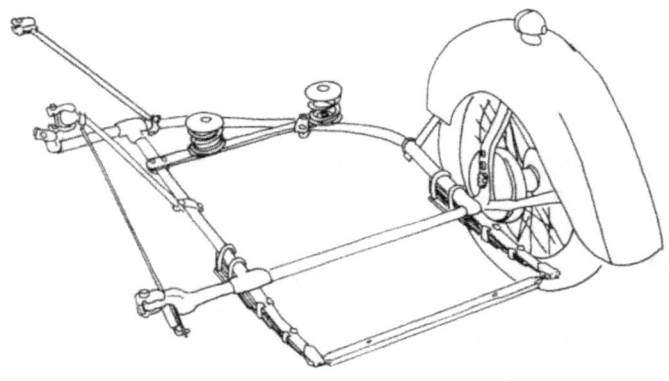

Version RB is sprung with quarter elliptical leaf springs (the half of a half-elliptical blade spring is quarter elliptical)

Supporting the front, there are two compression springs, each one fitted between spring cups, one end is fitted on a cross rail, the other half to the base of the body.

The backward pointing blade springs have either five or eight steel blades, clamped rigidly under the chassis frame tubes. Consequently, there is no need for links, (like with version R); the body rear cross bar can be placed directly on the mountings on the extremes of the springs.

Version RB made it possible to fit a wider box than on version R. Therefore at the introduction, version RB was recommended for pick-ups.

Chassis version R, no. 1201 – 2540 (version '38)
Tube frame with brake. The connections are made with lugs, instead of the tubes being butt-welded (flush) together.

On the foremost connector, a special shaped lug is fitted, so that the ball is positioned 40 mm above the plane of the chassis frame. The brake operating axle runs through the rear tube reinforcement and is rotated by the brake lever.
When the foot brake is operated, the sidecar brake lever is moved by a pull rod, together with the rear brake of the motorcycle. A 180 mm drum brake is fitted on the tube chassis, with the same construction as the brake of the motorcycle's rear wheel. With the adjustment mechanism, the two brakes can be adjusted and coordinated such that the vehicle does not pull to one side on braking.
This application was a considerable improvement of the motorcycle's braking ability.
With the introduction of this sidecar chassis, there were only three ball connectors. A fourth angled reinforcement rod was quickly introduced however; a 492 mm long straight steel rod with a ball.

Chassis version RB, no. 2541 – 3424 (version '48)
The construction of this tube chassis is without lugs and the front connectors are bent 40 mm up in relation to the plane of the chassis frame. Both angled connectors are as version '38. The rear springing consists of two quarter elliptical springs. The springs for the passenger body has only four steel blades, that for a pickup eight. The rear springs are each fitted with a 25 mm eye, instead of a screw-on link. He rear cross rod rests in a pair of rubber bushings which are pressed into the two eyes. The brake rod runs through a rubber bushing.

-*From 2613* the cross rails for the two front compression springs were made wider and stronger.

-*From 2764* the sidecar brake with self-centering brake drums is fitted, the same as with the brakes of the motorcycle. But the factory continued with fitting pressed brake drums after the cast brake drums with rib(s) were standard for the rear wheel.

Chassis, version RB, no.3425 – 6194 (version '50)
As version '48, but the eye on the springs are 30 mm; the rubber

bushing was changed to 30mm and the rear support rod has discs to stabilize the rubber bushings. From the same serial number (3425), an improved adjusting mechanism for the brakes was fitted (see below).

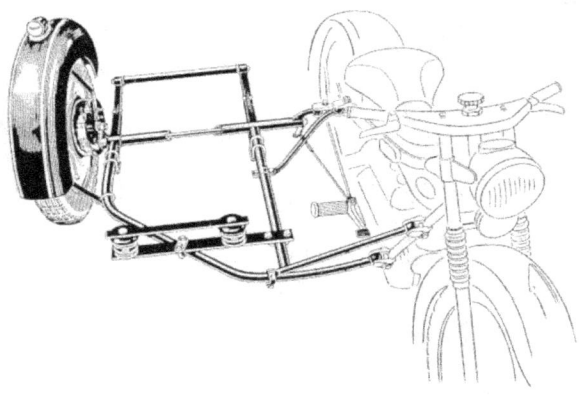

-From appr. 4100 (corresponding with Nimbus no. 10440) the front angled rod is extended by 60 mm to 552 mm. This allows the clamping device of the top front ball joint to be detached from the cross reinforcement. This will reduce the chance of the frame tearing as a result of overstressing, and thus causing an accident.

F&N

BRAKE OPERATION

The operating mechanism for the sidecar brake is fitted as a unit with the sidecar brake pull rod as follows: The pull rod is secured with a washer with split pin in the hole at the rear of the brake pedal, and the hook of the adjustment mechanism is secured with a washer and split pin in the lever of the sidecar brake:

-R 1201 – R 1400 were fitted with a(n) (unknown) construction, that was too weak.

-R 1401 – RB 2620 were fitted with a more robust construction.

-RB 2661 – more robust; fitted to all later sidecar chassis.

F&N

SIDECAR WHEEL

All wheels for Nimbus sidecars have 3" x 19" steel rim with 40 holes.
These rims can be enameled or chrome plated (or, later, made from stainless steel).

The chassis of the version '34, 501 – 565 are fitted with a regular steel tube hub with two ball bearings.

The chassis U 701 – U 1025 are fitted with a malleable iron hub and with separate bearing races with 13 pieces of 5/16" balls inside the chassis frame and 12 pieces in the hub cap.

From chassis R 1201 the hub is made from malleable iron with a flange for the screwed-on brake drum and internal recesses for fitting two conical roller bearings.

Brake drums

On sidecars with brake, the same brake drum is used as on the rear wheel of the motorcycle.

However, after malleable iron brake drums were introduced for the motorcycle (1950), pressed steel brake drums continued to be used on sidecars.

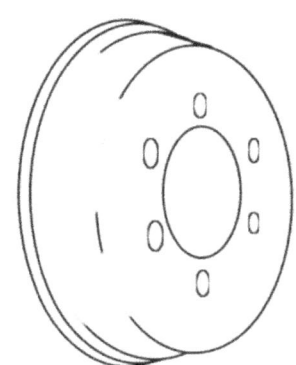

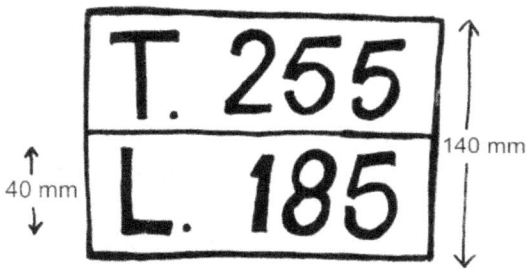

Total weight (T) and *max. cargo* (L)

Hub cap

The hub cap is made from aluminium; it protects the bearing from water and dirt.

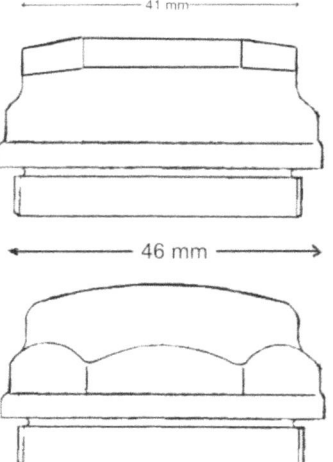

There are two versions:
-501 – 565 and U 701 – U 1025 with a 41 mm hex at the extended end.

-R 1201 – RB 6194 with a 46 mm hex directly on the threaded section.

SIDECAR MUDGUARDS

The mudguard for the Nimbus sidecar chassis is made from steel with a side bracket protruding into the body.

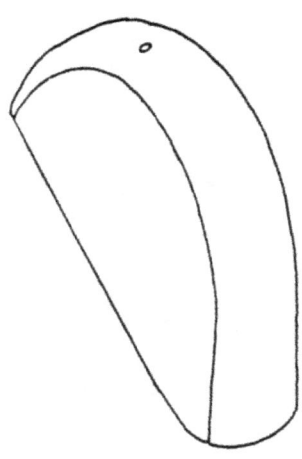

Sidecar mudguard 501 -565 and U 701 – U 1025
The mudguard is closed at the side towards the sidecar body.

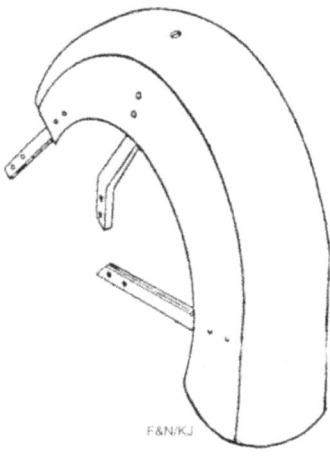

Sidecar mud screen R 1201 – RB 3424
The mudguard is partly closed at the side towards the sidecar body and has a rear end profile, exactly as with the rear mudguard of the motorcycle (1301 – 9000)

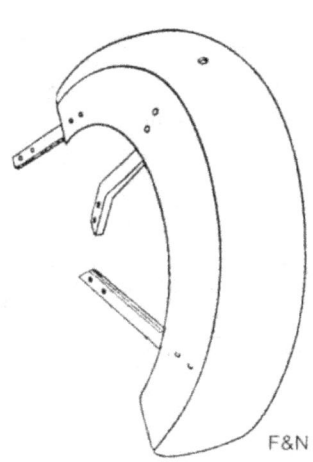

Sidecar mud screen RB 3425 – RB 6194
The mudguard is partly closed at the side towards the sidecar body and a rear end profile, with an arch which is the same as with the rear mudguard of the motorcycle (9001 - 14015)

SIDECAR LIGHTS

The light for Nimbus-C sidecar is fitted on the mudguard and is an integrated head- and taillight.
Different makes of lights were used. Three of these are known:

Willoq-Bottin
Light with a forward pointing large white light orifice and a smaller backward pointing red "cat's eye", both lit by the same light bulb. The lamp housing can either be enameled in the color of the mudguard, or chrome plated.

Brenner
Light with a forward pointing small white orifice and a rather insignificant red opening, which glares backward. The lamp housing can either be enameled in the color of the mudguard, or be chrome plated.

After May 24, 1955, both aforementioned lights became illegal. A/S Fisker & Nielsen planned to produce a Nimbus side car light complying with the new requirements. They made drawings for it, but it was never produced. The light mentioned below was fitted instead:

Ermax
Ermax is a standard light, fitted onto numerous vehicles as a parking light, indicator light etc. It has a housing from black plastic, a white forward shining light and an equally large red beam shining backward.
The red glass with built-in reflector is marked "J.R.U", which means that this type of light is approved by *Justitsministeriets Refleks Udvalg* (Reflector Commission of the Ministry of Justice). The light is not enameled, but can be fitted with a chrome plated cover shield that is a snap-on fit.

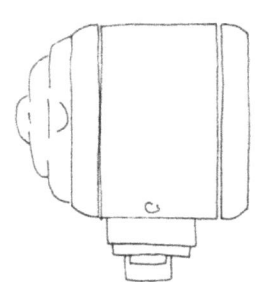

OVERVIEW – SIDECARS

Drawings and text can give us a broad overview about the technical development of sidecars. In addition, the actual sales prices of new Nimbus sidecars are listed over the whole production period.

In the description, the author tried to be more consistent with respect to the terminology, than the factory did.

The brochures, prospectuses and price lists form the base of this overview. The listed sales prices are generally the sales prices for the customer, including VAT and registration costs, but excluded road tax and insurance.

The overview cannot stand by itself and has to be read in conjunction with the other information in this book. The list of key words may also be helpful to find the subject you are looking for.

1934

Version:	1934 (501 - 550)
Chassis:	Version '34; tube chassis, version '34 without brake.
Bodies:	Standard passenger sidecar with adjustable wind screen.
	Utility sidecar with box and lid. Internal dimensions of box: 1400 mm long, 500 mm wide, 500 mm high.
Color.	Chassis, mudguard with light, wheel hub, spokes, rim and box: black with no pinstriping.
Price:	'Standard' sidecar: DKK 400.-. Utility sidecar: DKK 425.-.
Number:	Stamped in the tube chassis near the front connector.

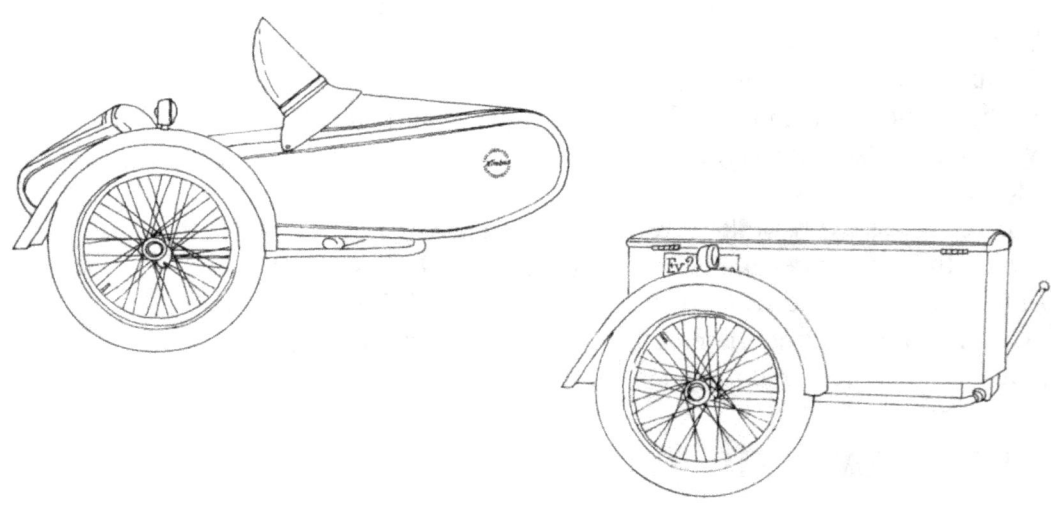

1935

Version:	1935 (551 - 565 and U 701 - U 814)
Chassis:	551-565: Version '34, tube frame with no brake. U 701- U 814: Version U, Version '35, flat steel chassis with no brake.
Bodies:	1) Passenger body with adjustable windscreen with no door (on the version '34 chassis) 2) Passenger body with adjustable windscreen and door (on chassis version U). 3) 'Luxus' passenger body with adjustable windscreen and door (on chassis version U). 4) Utility sidecar box; internal box dimensions: 1400 mm long, 500 mm wide, 500 mm high.
Colours:	All chassis: black. 1), 2) and 4): all parts black, without pinstriping. 3) chrome plated rim and spokes; mudguard and box in the colour and with pinstriping as the motorcycle, which means red or green with single gold pinstriping, light chrome plated.
Prices:	Chassis with 1) DKK 425.- Chassis with 2) DKK 500.- Chassis with 3) DKK 545.- Chassis with 4) DKK 450.-. Chassis with wheel, mudguard etc. but excluding the box: DKK 250.-.
Number:	Stamped in the chassis near the front connector.

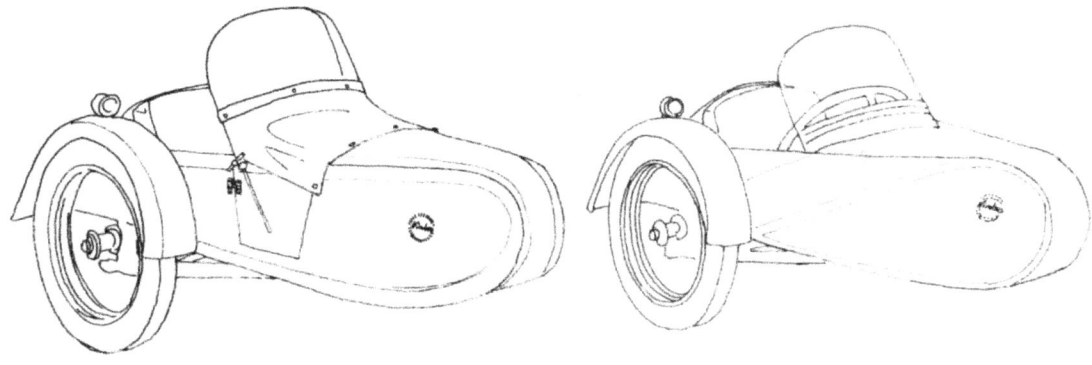

1936

Version:	1936 (U 815 - U 948)
Chassis:	Version U, version '35, flat steel frame without brake.
Bodies:	1) 'Standard': Passenger body with adjustable windscreen, without door.
2) 'Luxus': Passenger body with adjustable windscreen, with door.
3) 'Standard' utility sidecar box: Internal box dimensions: 1350 mm long, 500 mm wide, 500 mm high.
Colours:	All chassis: black.
'Standard': wheel, mudguard and sidecar light: black with no pinstriping.
'Luxus': mudguard with light and box: red or green with single pinstriping; wheel aluminium bronze enameled.
Prices:	1) 'Standard' passenger sidecar DKK 500,-
2) 'Luxus' passenger sidecar DKK 525,-
3) 'Standard' utility sidecar DKK 450,-
4) 'Standard' flat steel chassis with wheel, mudguard and light, but without box: DKK 250,-.
Number:	Stamped in the chassis near the front connector.

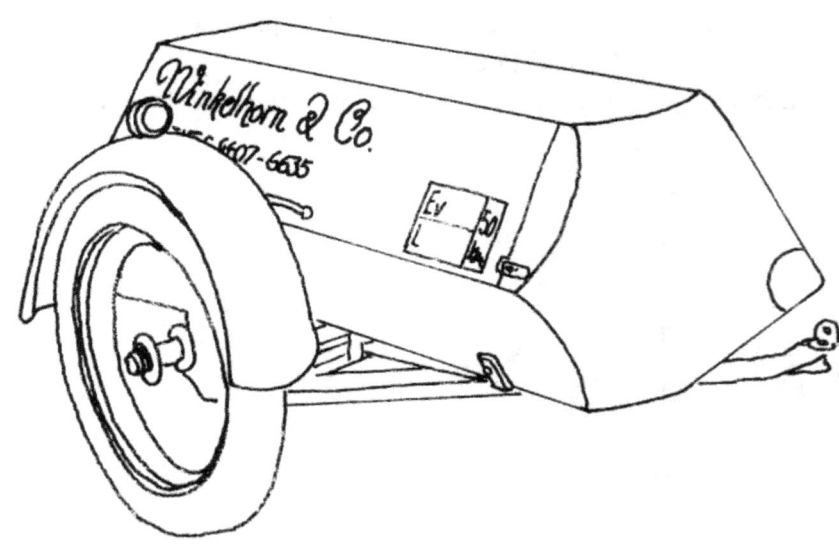

1937

Version: 1937 (U 949 - U 1025 and R 1201 - R 1240)

Chassis: U 949 - U 1025: Version U, version '35, flat steel chassis with no brake.

R 1201 - R 1240: Version R, version '38, tube steel frame with brake.

Bodies: 1) 'Standard': Passenger body with adjustable windscreen, without door, (fitted on chassis version U).

2) 'Luxus': Passenger body with door (fitted on chassis version U).

3) 'Sport': Passenger body with non-adjustable windscreen and no door (on chassis version U or R).

4) 'Standard' utility box internal dimensions: 1350 mm long, 500 mm wide, 500 mm high (fitted on chassis version U).

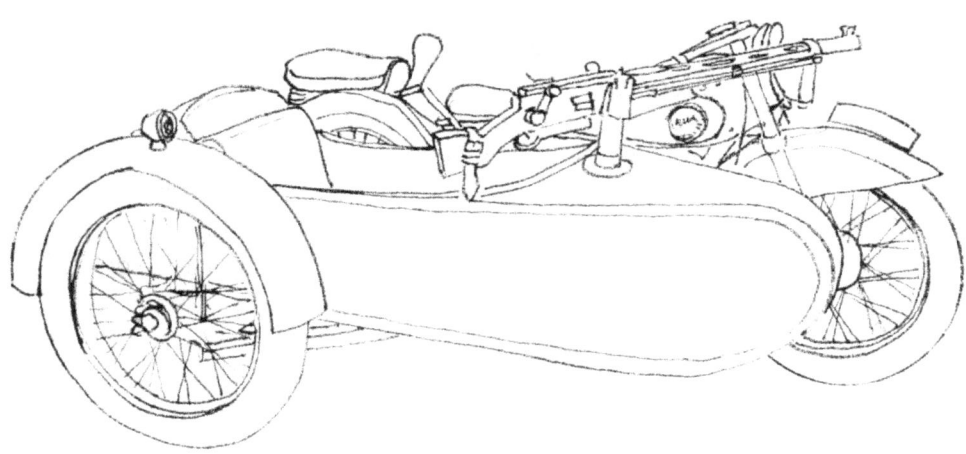

Colors: All chassis: black.

1) 'Standard': body, wheel, mudguard, and sidecar light: black with no pinstriping.

2) 'Luxus': red or green with single gold pinstriping; wheel: aluminium bronze enameled.

3) 'Sport': body and mudguard: blue with single silver pinstriping, sidecar light chrome plated, wheel with aluminium bronze enameled hub, chrome plated rim and cadmium plated spokes.

Prices:	1) 'Standard' passenger sidecar DKK 500,- 2) 'Luxus' passenger sidecar DKK 525,- 3a) 'Sport' passenger sidecar on chassis version U: DKK 525,- 3b) 'Sport' passenger sidecar on chassis version R: DKK 710,- 4) 'Standard' utility sidecar DKK 450,- 'Standard' chassis (version U) with wheel and mudguard with light: DKK 250,-
Number:	Version U: stamped in the frame near the front connector. Version R: on an identification plate, riveted onto the left hand clamping plate in front of the blade springs.

1938

Version:	1938 (R 1241 - R 1403)
Chassis:	Version R, version '38, tube frame with lugs, brake fitted. Front angled rod 492 mm.
Bodies:	1 + 2) 'Standard' and 'Luxus': Passenger body with adjustable windscreen, and no door. 3) 'Sport': Passenger body with non-adjustable windscreen, without door. Wheel with aluminium bronze hub, chrome plated rim and cadmium plated spokes. 4) 'Standard' utility box, internal dimensions for box: 1350 mm long, 500 mm wide, 500 mm high.

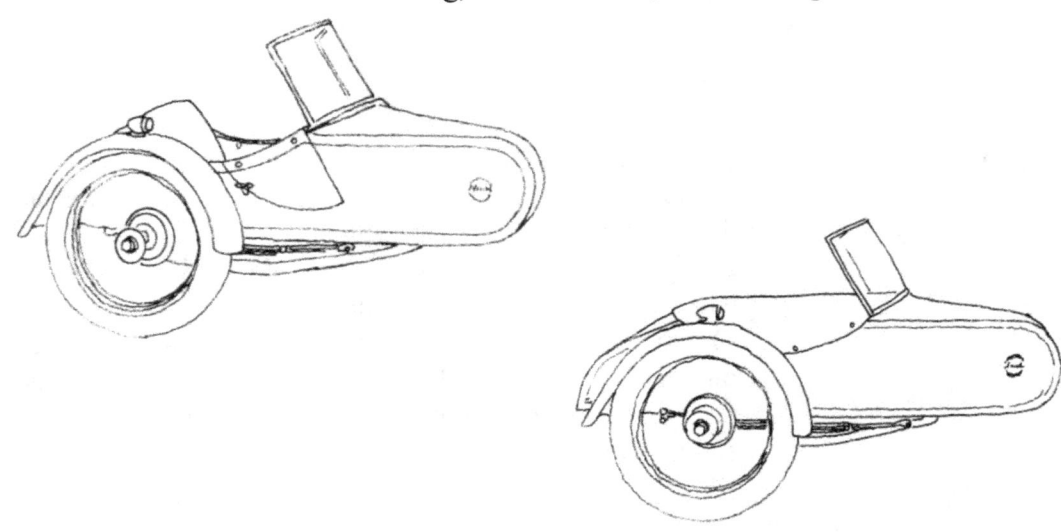

Colours:	All chassis: black. 1 + 2) 'Standard' and 'Luxus': black, red or green with single gold pinstriping; wheel: aluminium bronze enameled. 3) 'Sport': Body and mudguard: blue with single silver pinstriping, sidecar light chrome plated, wheel with aluminium bronze hub, chrome plated rim and cadmium plated spokes.
Prices:	1 + 2) 'Standard' and 'Luxus' passenger sidecar: DKK 700,- 3) 'Sport' passenger sidecar DKK 710,- 4) 'Standard' utility sidecar DKK 590,- 'Standard' chassis, version R; wheel, mudguard and light: DKK 350,-
Number:	On an identification plate, riveted onto the left hand clamping plate in front of the blade springs.

1939

Version:	1939 (R/RB 1404 - R/RB 1800)
Chassis:	Version R/RB, version '38, tube frames, assembled with lugs, brake fitted. (Both types of springing are produced. See identification plate, registration document or stock book) Front angled bar 492 mm.
Bodies:	1 + 2) 'Standard' and 'Luxus': Passenger body with adjustable or non-adjustable windscreen. 3a + 3b) 'Sport' and 'Special': Passenger sidecar with non-adjustable windscreen Wheel with aluminium bronze hub, chrome plated rim and cadmium plated spokes. 4) 'Standard' utility sidecar box internal dimensions: 1350 mm long, 500 mm wide, 500 mm high.

5) 'Ambulance': Bottom body steel, to fit the Danish 'Standard' stretcher with tarpaulin cover.

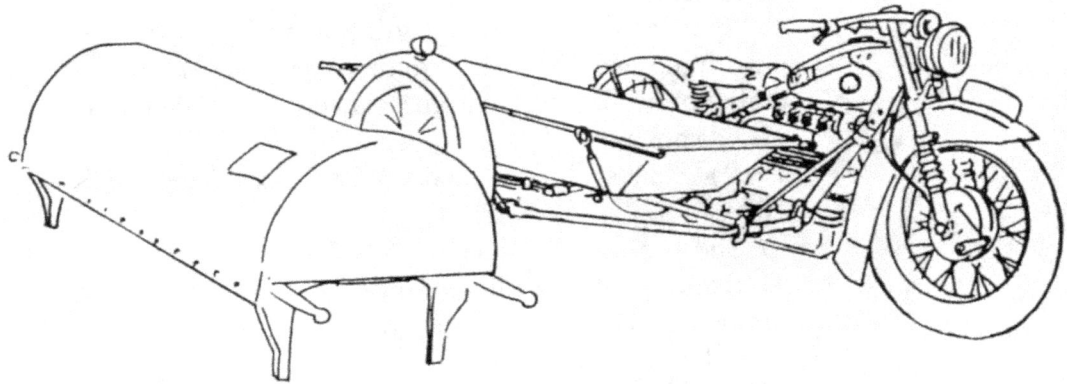

Colours:	All chassis: black. 1 + 2) 'Standard' and 'Luxus': Body and mudguard black, red or green with single gold pinstriping; wheel, aluminium bronze enameled. 3a + 3b) 'Sport' and 'Special': Body and mudguard blue with single silver pinstriping or ivory yellow or lavender/grey single gold pinstriping, sidecar light chrome plated, wheel with aluminium bronze hub, chrome plated rim and cadmium plated spokes.
Prices:	1 + 2) 'Standard' and 'Luxus' passenger sidecars DKK 762,50 3a + 3b) 'Sport' and 'Special' passenger sidecars DKK 774,- 4) 'Standard' utility sidecar: DKK 565,- 5) Stretcher sidecar without stretcher: DKK 300,- 'Standard' chassis, version R or version RB with wheel, mudguard and light: DKK 325,-
Number:	The identification plate for version R is riveted onto the left hand clamping plate in front of the blade springs and for version RB onto the tube chassis between the clamping brackets for the front springs at the left hand side.

1940-46

Version: 1940-46 (R/RB 1801 - R/RB 2425)

Chassis: Version R/RB, version. '38, tube frame assembled with lugs, brake fitted. (Both types of springs are produced. See identification plate, registration document or stock book.) Front angled bar 492 mm

Bodies: 1 + 2) 'Standard' and 'Luxus': Passenger body with adjustable or non-adjustable windscreen.

3a + 3b) 'Sport' and 'Special': Passenger body with non-adjustable windscreen. Wheel with enameled hub, chrome plated rim and cadmium plated spokes.

4a) Utility box version R, box internal dimensions: 1350mm long, 500 mm wide, 500 mm high.

4b) Utility box, type RB, box internal dimensions: 1150 mm long, 600 mm wide 525 mm high.

5) 'Ambulance': Bottom body steel, to fit the Danish 'Standard' stretcher with tarpaulin cover.

Colours: All chassis: Black

1 + 2) 'Standard' and 'Luxus': black, red or green with single gold pinstriping; wheel: aluminium bronze enameled.

3 a + 3 b) 'Sport' and 'Special': blue with single silver pinstriping or ivory/yellow and lavender/ grey with single gold pinstriping, sidecar light chrome plated, wheel with aluminium bronze hub, chrome plated rim and cadmium plated spokes.

Prices 1940:	1 + 2 + 3) Passenger sidecar box: DKK 375,- 4a) Utility sidecar, box dimensions 1350 mm x 500 mm x 520 mm: DKK 275,- 4b) Utility sidecar, box dimensions 1150 mm x 600 mm x 520 mm: DKK 275,- 5) Stretcher sidecar without stretcher: DKK 300,- Chassis for passenger sidecar DKK 360,- Chassis with mudguard, light and wheel for utility sidecar version R or RB: DKK 375,-
Prices 1942:	1 + 2) 'Standard' and 'Luxus' passenger sidecar: DKK 910,- 3a + 3b) 'Sport' and 'Special' passenger sidecar: DKK 920,-
Prices 1943:	1 + 2 + 3) Chassis version R: DKK 420,- (passenger); 4) DKK 432,- (Utility). Chassis version RB: DKK 450,- Utility box: DKK 330,-
Number:	The identification plate for version R is riveted at the left hand clamping plate in front of the blade springs and for version RB on the tube chassis between the clamping brackets at the left hand side.
Note:	The sales of passenger sidecars stopped during 1943 because of the occupation production priorities.

1947

Version:	1947 (R/RB 2426 - R/RB 2618)
Chassis :	Version R/RB, version '38, tube chassis assembled with lugs, with brake. (Both springing versions were produced, see identification plate, registration document or stock book). Front angled rod 492 mm.
Body:	No bodies; the factory supplied only sidecar chassis with wheel, mudguard and light.
Colours:	All chassis: black. Mudguard: black or upon request in the coulour of the motorcycle, including pinstriping. Wheel with aluminium bronze hub, rim and spokes.
Price:	Chassis for passenger- and utility sidecar DKK 645,- plus DKK 36,- delivery cost.
Number:	The identification plate for version R is riveted onto the left hand clamping plate in front of the blade springs

and for version RB on the tube frame between the clamping brackets for the blade springs at the left hand side.

1948 - 51

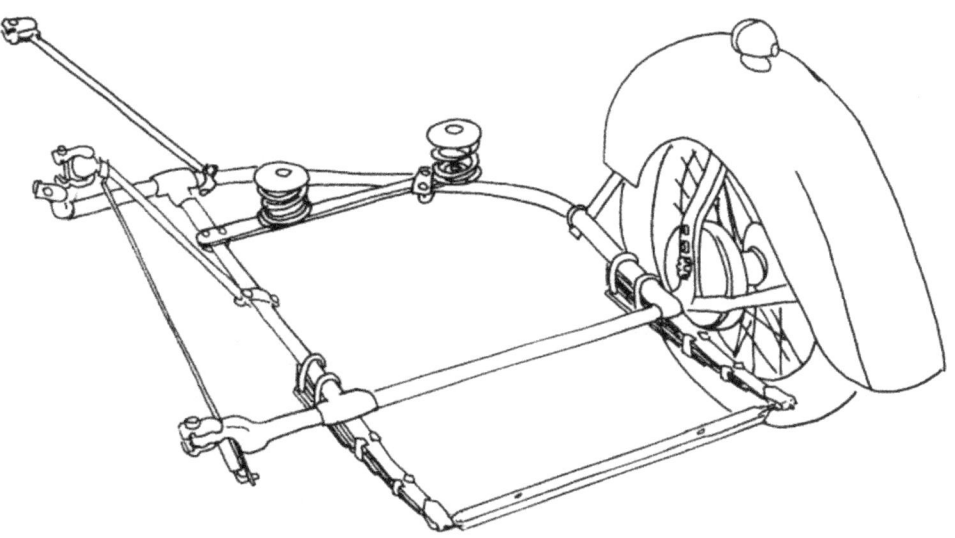

Version:	1948 - 51 (RB 2619 - RB 4099)
Chassis:	Version RB, version '48/50, tube frame assembled without lugs, brake fitted. The front clamping arm is bent up. The front angled rod is 492 mm.
Body:	No bodies; the factory supplied only chassis with wheel, mudguard and light.
Colours:	All chassis: black. Mudguard: black or, upon request, in the colour of the motorcycle, with pinstriping. Wheel with aluminium bronze hub, rim and spokes.
Price:	Passenger - and utility chassis DKK 645,-
Number:	Identification plate, riveted onto the tube chassis between the clamping brackets for the blade springs, at the left hand side.

1952 -53

Version:	1952 -53 (RB 4100 - RB 5106)
Chassis:	Type RB, version '50, tube frame, assembled with lugs,

	brake fitted. The front clamping arm is bent up. Front angled rod 552 mm.
Body:	No bodies. The factory supplied only sidecar chassis with wheel, mudguard and light.
Colours:	All chassis: black.
	Mudguard: black or, upon request, the colour of the motorcycle. Wheel with aluminium bronze hub, rim and spokes.
Price:	Passenger- and utility chassis DKK 720,-
Number:	Identification plate, riveted onto the tube chassis between the clamping brackets for the blade springs, at the left hand side.

1954 - 56

Version:	1954 - 56 (RB 5107 - RB 5876)
Chassis:	Version RB, version '50, tube frame, assembled without lugs, with brake. The front clamping arm bent up. Front angled rod 552 mm.
Body:	Passenger- and utility body upon request only.
Coulors:	All chassis: black.
	'Standard': mudguard black, wheel: aluminium bronze enameled.
	'Luxus': mudguard: black, red withered/green or deep sea green with double gold pinstriping; Rim, spokes and nipples chrome plated.

Prices: January 1st, 1954: passenger - and utility chassis:
'Standard': DKK 700,-
'Luxus': DKK 720,-
April 1st, 1954: Passenger- and utility chassis:
'Standard': DKK 690,-
'Luxus': DKK 700,-
Utility vehicles with yellow number plates: no VAT, with black/yellow number plates half VAT (See prices)

Number: Till March 31, 1956: on the identification plate between the clamping brackets for the blade springs at the left hand side. From April 1st, 1956: version and number pressed in the tube frame near the rear lower clamp.

1957 - 59

Version: 1957 - 59 (RB 5877 - RB 6194)

Chassis: Version RB, version '50, tube frame assembled without lugs, brake fitted. Front clamps bent up. Front angled rod 552 mm

Body: Passenger – and utility bodies only upon request.

Colours: All chassis: black
'Standard': mudguard black with double gold pinstriping; wheel: aluminium bronze enameled.
'Luxus': mudguard black, red, withered green or deep sea blue with double gold pinstriping; rim, spokes and nipples chrome plated.

Prices: Passenger- and utility chassis:
'Standard': DKK 690,-
'Luxus': DKK 700,-
Utility vehicles with yellow number plates, no VAT, with black/yellow number plates half VAT. (See prices).

Number: Version and number are stamped in the tube frame near the rear, lower clamp.

MATERIALS AND TECHNIQUES

MATERIALS

Nimbus-C was designed by father and son P.A. Fisker and Anders Fisker. The actual engineering work, including the calculations, choice of materials and processing, was solely done in 1932-33 by Anders Fisker however. It is beyond the scope of this book to list all materials for all parts used in the construction of this motorcycle. By reading through the various chapters of this book about assemblies and components, most of the materials are dealt with. But to give an impression of the many materials used, please find below a summary of them, quoted from Anders Fisker's own notes.

Cast iron
Malleable cast iron
Aluminium, Gun metal, Brass, Bronze - cast
Forged steel 0.20%, 0.30%, 0.50%
Special steel FR 86 – HR 33 – SR 185
Stainless steel – length 5 meter
Round steel: 0.35%, 0.20%, Auto
Hex steel: 0.35%, Auto
Tube: steel-, Copper-, Aluminium-, Brass-
Round and hex Brass
Phosphor bronze, Springs bronze, Silver, White metal
Copper fins
Flat steel 0.35 % (six different dimensions)
Steel plate, 3.5 LHK (five different dimensions)
Brass plate (seven different dimensions)
Gaskets: Fiber, Insulating pressed paper, Leather, Cork, Felt
Cables and Leads: Bowden-, Ignition-, Low voltage, Flex
Springs
Bearings: Slide bearings, Ball bearings, roller bearings, Thrust bearings, Balls
Nokait, Bakelite, Ebonite, Rubber, Friction materials
Split pins, Keys, Rivets, Washers, Spring washers

Please note, that steel plate, flat steel and other types of steel were in the early thirties often stamped with the trade mark of the Swedish steel company Fagersta AB, a swastika. When sandblasting a frame, seat or pillion seat, sometimes this symbol can be seen, a few centimeters long, stamped on the surface of the material. This has nothing to do with Nazism or occupation! (By the way, Carlsberg were using this symbol during the same period).

BOUGHT-IN PARTS

Not all parts for Nimbus-C were produced by A/S Fisker & Nielsen. When the 'Bumblebee' was taken into production in 1934, they acted like all modern industries; subcontractors and suppliers were selected for a long range of parts for this motor cycle.

It would have been both unprofitable and inappropriate to develop in-house highly specialised parts, as mentioned below.

Yet there were actually many parts that the factory produced themselves. Most "motorcycle factories" in those days bought a frame here, an engine there and wheels and mud guards at another place, and were in fact only an assembly plant.

The next overview of bought-in parts comes from Anders Fiskers own notes from January and August 1934, as well as from January 1935.

Front mud guard (A.Höfer)
Rear mud guard (A.Höfer)
Fuel tank (Glud and Marstrand)
Rims (Dunlop W.M. 3-19)
Spokes (Taylor)
Nipples (Taylor)
Cylinder head gasket – Cu – Asbestos
Exhaust gasket - ditto
Head light (Riemann)
Horn (Riemann)
Head light bracket
Horn push button (Bosch SSH 506/lz)
Rotor (Bosch ZVT 23/lz)
Spark plug (Lodge HD 14)

Battery (Noack)
Contact breaker
Condenser (Hydra Werke)
Ammeter (1934: Schoeller, 1935 VDO)
Speedometer (VDO)
Speedometer cable (VDO)
Bilux light bulb, 15/20W, 20 mm
Light bulb, 3 W
Fuel valve (Zöblitz)
Nipple and union nut for fuel pipe
Fuse – 30 mm – 20 A
Float - carburettor
Piston – alloy 280 (K.Schmidt)
Piston ring (A.Tevis)

Gudgeon pin with plugs (Wizemann)
Valve GSE (A.Tevis)

Carbon brush dynamo, 6 x 12 and 3 x 12
Carbon brush, distributor
Fibre gasket, Dynamo
Name plate
Decals
Red celluloid plate, Taillight
Blank celluloid plate, Taillight
Grease nipple AM 6 x 0.75 DIN 3402
Cable bracket, small
Cable bracket, large
Air pump with hose (Bluemel's)
in the Toolbox:
- Grease gun
- Angled tyre lever
- Straight tyre lever
- Combination pliers 7"
- Screwdriver
- Adjustable spanners 8"
- Spark plug spanner
- Box spanner 10 – 14 mm
- Open end spanner 6 mm with feeler gauge
- Same. 10 - 14 mm
- Same 17 – 19 mm

Cotter pin for dynamo shaft top and bottom
Slit pin for conrod
Split pin for clutch
Split pin for lube oil pump spring
Split pin for brake rod
Split pin for Bowden cable clutch -brake
Split pin for carburettor tickler
Split pin for brake shoe
Key for flywheel (Woodruff)
Conical pin – kick start axle
Pin 8 mm – Engine block
Pin 5 mm - Gearbox
Pin 4 mm – Front and rear axle
Bowden cable for brake
Bowden cable for clutch
Bowden cable for carburettor
Cork gasket – Fuel cap
Cork gasket for gearbox - Out
Cork gasket for gearbox - In
Cork gasket for oil filling
Cork gasket for kickstarter axle
Leather gasket – oil drain
Fibre gasket – dynamo neck
Insulation - distributor
Cork gasket – clutch release bearing
Sleeve – front fork

SURFACE TREATMENT

In notes from 1934, Anders Fisker enumerated how the different parts had to be surface treated. Many of these surface treatments became obsolete during the production period 1934 – 59 and were substituted by other techniques. Other treatments were no longer allowed because of environmental implications.

As is the case with the applied materials, it will go too far to list each and every part and how it was treated. In reading through the text of the various chapters, surface treatments of most parts are dealt with.

In the following text the original and some of the alternative surface treatment processes are briefly explained.

All surface treatment processes have their impact on the environment and the majority of those must absolutely not be done at home!

Parkerizing

Parts from iron or steel, carefully cleaned from rust and oil, are carefully submerged in a solution of iron-, zinc- and manganese phosphates. After 60 – 70 minutes, the surface is covered with a layer of light grey phosphate. This layer protects the iron from getting rusty again, but is very susceptible to grease and oil and will, when left unprotected, soon become contaminated. Parkerizing builds a base for further enameling or oil treatment.

Bonderizing

Bonderizing is in broad lines the same as parkerizing, but instead of zinc phosphate, the bath contains warm copper salts. This makes it possible to reduce the processing time by about 10 minutes. The surface is not really rust resistant, but on the other hand is a good base for further enameling, to make the latter last longer.

Immersion enameling

In this process, parts are (repeatedly) submerged in thinned cellulose enamel. This process is especially suitable for small parts, that have previously been parkerized.

Oil treatment

Oil treatment is a surface treatment where parts are immersed in warm oil. This process is applied to parts which previously have been parkerized. This treatment must not be confused with hardening / tempering in an oil bath. The surface obtains a certain protection against the oxygen in the air.

Stove enameling

Stove enameling is a process of spraying cellulose enamel onto a well treated surface (see parkerizing or bonderizing). After enameling, the parts are hung and treated in an oven at 250° C to let the enamel harden during approximately 60 – 70 minutes, which makes the enamel hard and very dense. Often times a number of thin layers are applied, which are hardened after each new layer.

Nickel plating / matt nickel plating

In an electro galvanic plating process, a more noble metal layer is laid over a less noble metal part.
A process is called galvanic, when the part to be treated (base metal) acts as the cathode (minus) in an electrochemical bath, the anode (plus) being the more noble metal (in the galvanic series); the base metal part is covered with a layer of the more noble metal. When an electric potential of 2.5 – 4 Volt is applied, metal precipitation takes place with a certain current, defined in Amps / dm^2.
Nickel is a very suitable metal for parts to be coated by means of the electro galvanic process, but the surface has to be covered with copper (or silver) first. Therefore, the process preceding the nickel plating is galvanic copper plating. For the nickel plating process, a mixture of potassium cyanide and copper sulphate is used in the bath.

The most common method of galvanic nickel plating is matt nickel plating. If in this case shiny nickel plating is wanted, the part to be plated has to be ground smoothly and the nickel layer has to be polished.

Chrome plating

Chrome is a metal that can be used as a surface treatment, by using a galvanic process, comparable as with nickel. In 1934 – 35, little expertise was gained with respect to chrome plating and the process was relatively expensive, partly because the parts had to be copper- and nickel plated first, and partly because chrome was expensive. Last but not least, the process is controlled best in a rotating machine or drum. However, a correctly executed chrome plating process will result in a very resistant surface.

Nowadays, there are many places where chrome plating can be done at a reasonable price. But there is of course a relationship between quality and price!

Hard-chrome plating

A special, expensive electro galvanic technique makes it possible to apply a relatively thick layer of chrome to surfaces, that are subject to high forces. For NIMBUS-C successful tests have been done with hard-chrome plating of crankshaft journals.

Galvanizing

The term galvanizing is commonly used in those cases where a surface is covered with a thin layer of metallic zinc, hence, a galvanic process where zinc acts as the anode. Originally, this process was not applied on Nimbus parts. This method of surface treatment has been used over a long time however. The surface looks nice, the process is cheap and the layer is water- and moisture resistant. But the resistance against slush and road salt is extremely poor, so the application is not wide spread.

Cadmium plating

For small (threaded) parts, plating with a thin layer of cadmium is an effective and durable surface treatment. Cadmium is easy to precipitate and has an even light grey appearance. Like parkerizing (see above), the layer is durable and a good base for enamel coating. Cadmium however is a very poisonous heavy metal and because of serious environmental damage, it has been forbidden for many years.

Browning

Browning is a chemical surface treatment process where oxidation / rust formation is under control. By applying an oxidizer, mixed with a vegetable oil, a black/brownish surface is achieved, which is resistant to further deterioration. This method has been applied to Nimbus-C parts during the thirties.

REFERENCES

Andersen, Jens Bisbjerg (1996): *Nimbus model C - Teknik og historie.* Forlaget Notabene.
English translation:
http://www.geutskens.eu

Jørgensen, Knud (2012):
NIMBUS – Maintenance. English Edition. BoD – Books on Demand, DK

The Nimbus Owners Group UK: http://www.nimbus.veetopia.com

Danmarks Nimbus Touring: http://www.nimbus.dk.

Nimbus colours: http:/www.RAL-Colours.de

Prototype 1956 - 59

INDEX

Ammeter **176**
Approval **19**

Baffle plate **84**
Battery **168**
Bottom bearing halves **44**
Brakes **157**
Brake light switch **187**
Bud front fork **142**

Camshaft **116**
Camshaft housing **118**
Cap for
    flywheel housing **100**
Carbon brush **168**
Carburettor **135**
Carburettor needle **135**
Carrier **191**
Centre stand **88**
Charge warning light **177**
Chassis **194**
Chassis frame **194**
Choke **135**
Chinese eyes **106**
Clutch **114**
Clutch pedal **27 – 29 f.**
Clutch release **28 f - 133**
Colours **36 - 38**
Combination switch **175**
Condenser **172**
Connecting rod/conrod **109**
Crankcase ventilation **139**
Crankshaft **108**
Crown wheel **155**
Cross bracket **84**
Cylinder block **98**

Decals/transfers **96**
Declutching rod **133**
Dipstick **103**
Distributor **171**
Distributor housing **172**
Drive shaft **134**
Driver and pillion seat **90**
Dynamo **168**

Eccentric cam **153**
Electrical wiring **183 - 185**
Electrics **167**
Enameling **218 - 219**
Engine **97**
Engine block **98**
Engine number **41**
Exhaust **124**
Exhaust pipe **125**

Final drive housing **154**
Fish plate **84**
Fish tail **125**
Flat steel frame **84 - 195**
Float **135**
Flywheel **41 - 113**
Foot rest **89**
Foot-operated
    clutch release **133**
Fork tube **141**
Frame **84**
Frame number **20 – 41 - 87**
Frame plate **84**
Frame rail **84**
Front fork **141**
Front mudguard **161**
Front number plate **47**

Front wheel **149**
Fuel hose **94**
Fuel pipe **94**
Fuel tank **93**
Fuel tank cap **94**
Fuel valve **94**
Fuel slide valve **94**
Fuse **185**
Fuse holder **185**

Gasket ring **104**
Gearbox 15 - **127**
Gear lever 15 - **132**
Gear selector **15**
Gear shift **132**
Gudgeon pin **111**
Guide rail **16**

Hand grip **89**
Handlebars **145**
Head light **179**
Head gasket **104** - 216
Head tube **84** - 141
Heat shield **124-125**
Heavy gauge
 rubber band
Horn **177**
HT coil 15 - **170**
HT lead **173**
Hub 16 – **149** - **153**
Hub cap **198**

Identification plate 21 – 23 – **194**
Ignition coil 15 - **170**
Ignition key **176**
Ignition switch surround 30 - **146**
Instrument illumination **186**

Jiffy stand **191**

Kickstarter **122**
Kickstart lever **122**

Knee rubbers **89**

Lead bracket **173**
Light switch twist grip 11 - **145**
Lube oil pump **100**
Lube oil pipe **101**
Lube oil return pipe **104**
Lube oil system **100**

Make- and number plate **23** – **177** - **194**
Manifold 11 – 12 – 104 – **125**
Materials **215**
Mudguard **161** - **200**
Mudguard brace 90 - **163**
Mud flap **161**
Mud screen **200**

Nipple 141 – 149 – 153 - 216
Number plate **47**
Numbers **41**

Oil pan **122**
Oil pressure operated
 cut-out switch 52 – **104** - 168
Oil scraper ring **112**

Passenger body 77 – **196** - **203**
Patents **13**
Pillion seat **91**
Pinion gear **154**
Pinstriping **39** - **40**
Piston **110**
Piston ring **112**
Production number 8 – 21 – 36 - **31**
Pull rod 10 – 27 – **132** – 157 - 197

Ratio **156**
Rear drive **154**
Ratio identification **45**

Rear mudguard **163**
Rear number plate **47**
Rear wheel 16 - **153**
Rigid driveshaft **117**
Regulator **169**
Retaining ring **122**
Rim **148**
Ring shaped
      head gasket **104**
Rocker **120**
Rocker guide **121**
Rotor **172**
Rotating valve **18**

Seat 17 - **90**
Securing plate **172**
Sidecar **192**
Sidecar body **192**
Sidecar light **201**
Sidecar mudguard **200**
Sidecar number plate 49 - **50**
Sidecar wheel **198**
Side bags **191**
Sounding rod **103**
Spark(ing) plug **175**
Speedometer **164**
Spoke **148**
Spring cups **107**
Standard colours **37**
Stickers **96**
Stock books **24**
Surface treatment **218**

Tail light **180**
Tank **93**
Tank badge **96**
Tank cap **94**
Throttle twist grip 11 - **147**
Tickler **135** - 218
Tools 9 – **165**
Toolbox **89**

Transfers/decals **96**
Transmission **156**
Tyre **148** – 189 - 217

Utility body **212**

Valves **106**
Valve guide **106**
Valve housing **105** - 107
Valve spring **106**
Varnished fixed
      type transfer **96**
Water based
      slide transfer **96**
Weight 20– **42** – 113 - 199
Version 26 – 27 - **35**
Wheels **148**
Wind screen 191 - 202
Wiring **183** – 184 - 185

Year of production **23**
Yoke 141 – 177 - 186

www.ingramcontent.com/pod-product-compliance
Lightning Source LLC
Chambersburg PA
CBHW082325220526
45470CB00008B/2406